LAND WITHOUT JUSTICE

By Milovan Djilas

THE NEW CLASS

LAND WITHOUT JUSTICE

CONVERSATIONS WITH STALIN

MONTENEGRO

THE LEPER AND OTHER STORIES

NJEGOŠ

THE UNPERFECT SOCIETY: BEYOND THE NEW CLASS

UNDER THE COLORS

THE STONE AND THE VIOLETS

MEMOIR OF A REVOLUTIONARY

PARTS OF A LIFETIME

WARTIME

MILOVAN DJILAS **LAND**

WITHOUT JUSTICE

TRANSLATED BY MICHAEL BORO PETROVICH
INTRODUCTION AND NOTES BY WILLIAM JOVANOVICH

A HARVEST/HBJ BOOK

HARCOURT BRACE JOVANOVICH

NEW YORK AND LONDON

MAP BY HAROLD K. FAYE

To the memory of
ALEKSA N. DJILAS

INTRODUCTION

A curious incident took place when Milovan Djilas was
brought before the Belgrade district court in December of
1956 on the charge that, on the occasion of the Hungarian
uprising, he had published statements "slandering Yugoslavia." During the summary proceedings of that trial there
occurred a brief interchange between the chief judge and
the prisoner which at a distance seems strangely irrelevant.
In giving the defendant's personal history, the chief judge
described him as a Montenegrin. Djilas, who otherwise bore
the extravagant indictment and even the sentence of three
years' "strict imprisonment" without discernible emotion,
leapt to his feet. "I object," he declared. "The statement
should show that I am a Yugoslav."

Now, no more violently—or justly—proud people lives
on earth than the people of Montenegro, and Djilas is one
of them. But he knew well what Chief Judge Voislav
Janković could mean by identifying him, at that moment,
as a Montenegrin. He could imply that Djilas's fierce
heredity impelled him against the communist regime: that
his heresy was inevitable because it was inbred. Djilas had
only recently been vice-president of Yugoslavia and the
newly elected president of its one-party parliament, yet the
court could perhaps suggest that he was not now governed
as much by his political experiences as by his racial memories. Djilas's grandfather (and likely his father, too) was a
rebel against the princes of Montenegro long before there

vii

existed a Yugoslav nation. Defiance was nothing new in a Montenegrin.

Such speculation on a small, if undeniably characteristic, incident would no doubt appear fanciful to the Yugoslav officials. They do not normally dwell on the distinctiveness of the several peoples who make up the Yugoslav nation: Serbs (including Montenegrins), Croats, and Slovenes. Certainly they have reason not to emphasize the composite nature of Yugoslavia: its history between the two world wars is a painful record of opposition, on religious and political grounds, between peoples who actually share a common ancestry and a common language. (It likely contributed to their success that the four chief founders of Communist Yugoslavia were representatively—though not so by design—Josip Broz-Tito, by birth a Croat, Alexander Ranković, a Serb, Edvard Kardelj, a Slovene, and Milovan Djilas, a Montenegrin.) Yet even communists must accept that all men carry some mark of their origins and that none is wholly free from the memories of childhood, when legend and tradition and the hopes and fears of one's elders give form to one's private world. Djilas writes in the pages that follow, the biography of his family and of his own life to the age of eighteen: "the story of a family may be the portrait in miniature of a land."

During his forty-six years Milovan Djilas has been at different times a revolutionary, a soldier, a political leader—and always a writer. Before the war his writings as a journalist, as well as his communist activities, made him an enemy of the Yugoslav royal government. Many years later, in January of 1954, a series of articles led to his expulsion from the Communist party of Yugoslavia and dismissal from high offices in the government. He had undertaken to write a philosophical defense of Yugoslavia's "national" communism after Tito broke with Stalin in 1948, but within a few years, to the dismay of Tito, Djilas went beyond official duty to criticize his own country's party.

Following his dismissal, he passed a year under close police watch, jobless and alone but still writing, before he was brought to trial—the first of three trials—in January of 1955. The charge was one of "hostile propaganda," arising from an interview he gave to the New York *Times*. Released on a suspended sentence, he returned to the small Belgrade apartment where he lived with his wife and small son, and there devoted the next two years to writing *The New Class* and *Land Without Justice*.

The manuscripts of both books were completed before his imprisonment at Sremska Mitrovica, fifty miles from Belgrade, in December of 1956, following the second trial, at which he was accused of "slandering Yugoslavia" in statements made to the French press and in an article he wrote for the American journal *The New Leader*. During October of 1957 he left his prison cell only long enough to be tried and sentenced a third time. Now the charge was one of disseminating opinions "hostile to the people and the state of Yugoslavia" through the publication in the United States of *The New Class*, a book describing communist leaders in the Soviet Union and Yugoslavia, mainly, as comprising a new class of bureaucrats, or oligarchs, who use power solely for their own ends.

"The revolution," writes Djilas in *Land Without Justice*, "gave me everything—except what I had idealistically expected from it." And what had he expected? The years of his manhood gave him the grim test of heroism—during the German and Italian occupation of his country he faced death again and again in three years of guerrilla warfare against the enemy—and these years had made of him both a respected intellectual and a feared party leader. But his restless mind and uneasy spirit yearned for something more. "So it has always been here: one fights to achieve sacred dreams, and plunders and lays waste along the way—to live in misery, in pain and death, but in one's thoughts to travel far." He sought, and could not find in these times, the simple beauty of human justice. Perhaps, his search must carry him back to another and earlier time, when he was a boy and youth in Montenegro, back to his origins.

There is but a thin line between the heroic and the grotesque, as between tragedy and comedy, and mankind has always viewed the few truly heroic peoples of the world with humor as well as respect. The legend of Montenegro offers to us both prospects. In part, it reflects Gladstone's testament that "the traditions of Montenegro exceed in glory those of Marathon and Thermopylae and all the war traditions of the world," and Lord Tennyson's conclusion, in his sonnet "Montenegro"—

> O smallest among peoples! rough rock-throne
> Of Freedom! Warriors beating back the swarm
> Of Turkish Islam for five hundred years,
> Great Tzernagora! never since thine own
> Black ridges drew the cloud and brake the storm
> Has breathed a race of mightier mountaineers.

In part, also, it reflects the many stories told, whether in malice or in delight, of the Montenegrins' fondness for self-glorification and their instinct toward excess. In the days of their princedom—its boundaries scarcely seventy-five miles square and embracing no more than a third of a million subjects—they spoke seriously of themselves and the Russians as comprising a force of "one hundred and sixty millions." After World War I, as inductees in the newly created Yugoslav army, some Montenegrins refused to follow the customary "count off" of soldiers in the line on the grounds that no Montenegrin could be expected publicly to declare himself "second" or "third." Instead, he might sing out "first after the first" or "first after the first after the first."

As conspirators the Montenegrins are no less individualistic, as Fitzroy Maclean suggests in his biography of Tito.

In 1940 Josip Broz-Tito, then secretary-general of the out-
lawed and hunted Communist party in Yugoslavia, sum-
moned party leaders to a secret conference at a house
outside of Zagreb. "All went well," writes Maclean, "except
that an overzealous lookout man opened fire on the Montenegrin delegation, who, with the careless enthusiasm of
their race, had mislaid their instructions." The lookout man
is mentioned no more, but somehow one senses that he, too,
was a Montenegrin.

The legend of Montenegro rests on history. One who
values a man's courage and a nation's freedom will recognize that the history of Montenegro, in the daring and the
suffering of its people, generation after generation, is un-
equaled in Europe. Serbian by race, Orthodox by faith,
Montenegrin by choice, they were, from the fifteenth to
the nineteenth centuries, the sole people in all the Balkan
peninsula who were never wholly subjugated by the Ottoman Turks or, subsequently, by any of the European
powers. While the Turks were besieging Vienna in 1529
and again in 1683, threatening central Europe, there remained at their backs a free Christian principality that
would never cease, despite the incredibly disproportionate
numbers opposite, to withstand them.

Coming from the north, Slavic tribes overran the old
Roman province of Illyria during the sixth and seventh
centuries; among them were clansmen who settled in the
small crater-like valleys lying amidst the barren mountains
bounded on the west by the Adriatic Sea and on the east
by the Macedonian plain. (From a distance the cold, gray
limestone ridges of Montenegro actually appear black,
hence, presumably, the name "black mountains"—*Crnagora*
in Serbian, *Montenegro* in Italian—although the name may
possibly be derived from an early ruling family, *Crnojević*.)
During the thirteenth and fourteenth centuries Montenegro
(then known as *Zeta*) was ruled by princes of the Nemanja
dynasty of Serbia. The Nemanja kings created a great

empire which at its zenith, under Tsar Stefan Dušan, stretched to the gates of Constantinople.

In June of 1389 on *Vidov Dan,* the day when "we shall see" what happens, tragedy befell the Serbs and their allies at the Battle of Kosovo. *Vidov Dan* is a date as familiar to every Serb as his own birthday, for it commemorates Christian valor in defeat—the sorest trial of courage—and it gave rise to a magnificent epic poetry woven about the legendary promise that one day again the Slavs would be free and united. After Kosovo many Serbians, still resisting, fled to the rocky fastness of Montenegro, there to found clans which were to become the curse, if also the glory, of that country until modern times. For it was these clansmen's fanatical pride in the exclusiveness of their families that led them to measure heroism as a vindication of one's name, thereby encouraging the blood feuds that—oftentimes no less than the constant threat of the Turks—caused Montenegrins until late in the nineteenth century to wear pistols as part of their daily dress.

A second *Vidov Dan* awaited the Montenegrins at the end of the seventeenth century. Following the last of the Crnojević ruling family, the land had been for two centuries governed—to the uncertain extent the clansmen would submit to agreement—by *vladike,* or prince-bishops, who were popularly elected and consecrated by the Serbian Orthodox Patriarch at Peć. But the Turks had made inroads into Montenegro. Cetinje, the capital, had three times been sacked; and mosques appeared in some villages as apostasy spread. In 1697 the clansmen gave to Vladika Danilo Petrović of Njegoši (the Petrović ancestral home) the right to choose his successor from his family, thus establishing a theocratic rule in which, since the bishops were unmarried, succession passed from uncle to nephew. The literary epic poem *Gorski Vijenac* or "Mountain Wreath," written by the foremost Serbian poet, Njegoš—who was, as Peter II, a later prince-bishop of Montenegro—describes the terrible decision confronting Vladika Danilo in sending the small Montenegrin force against those of his people who had under duress embraced Islam. The Cross was raised against the Crescent and a frightful slaughter followed. Montenegro was saved, as it had been wrought, in blood.

What had been true for four centuries, the independence and the importance of Montenegro, was at the end of the eighteenth century recognized by the powers of Europe. Under the greatest of the prince-bishops, Peter I, who ruled from 1782 to 1830, the Montenegrins allied themselves with the Russian tsar and with Serbia, which had in this period won its complete independence from the Turks. Led by Peter I and his successor, Peter II, the poet Njegoš (ruled 1830–1851), Montenegro carried the struggle relentlessly, year by year, to the Turks on all sides—to Bosnia and Hercegovina on the north and east, to South Serbia and Macedonia on the east and south, to Albania on the south. The cost of freedom was to continue great, not alone because men must pay for it with their lives, but also because men will differ over what freedom is.

It is at this period of his homeland's history that Milovan Djilas begins the account of his family's life in a land without justice, as he calls Montenegro, quoting Njegoš, who despaired often of bringing law and education to his *besudnja zemlja*, land without recourse. Njegoš, indeed, had reason to despair of the Djilas family in the person of Milovan's grandfather's uncle, the outlaw Marko, who flouted the prince-bishop's law against blood feuds. Marko was later murdered by a captain sent by Danilo II, the prince (no longer bishop) who succeeded Peter II and ruled until 1860. Aleksa Djilas, the grandfather of Milovan, avenged Marko's murder and fled to Nikšić, an area inhabited by Montenegrins but then under Turkish rule. From that time, after 1860, until the founding of Yugoslavia in 1918 the destiny of the Djilas family, as of all

Montenegrins, was in the hands of Prince (later King) Nikola, a cunning and capricious but also greatly talented autocrat who was a soldier, historian, and poet. After Aleksa Djilas returned to his home in "Old" Montenegro, at Župa, during the Serbian-Montenegrin war against Turkey in 1875–1878, he was murdered, probably with the cognizance of Prince Nikola. Aleksa was not avenged by his son, Nikola Djilas, but the son, too, was suspected of plots against the prince and imprisoned. When he pleaded his loyalty he was freed and later given land, as an officer in the Montenegrin army, at Kolašin on the border of Turkish territory. There his son Milovan grew to manhood.

Three more wars were to embroil the independent nation of Montenegro, and in each Milovan's father fought, as practically every Montenegrin male in every generation had fought. In 1912, when Milovan was a year old, his father went off to the First Balkan War, in 1913, to skirmishes following the Second Balkan War. In June of 1914, on *Vidov Dan*, a Serb patriot and conspirator assassinated Archduke Francis Ferdinand of Austria, setting off the Great War. That year and the next Djilas's father and uncles fought the encroaching Austrians and Germans, only to be betrayed, in Milovan's view, by King Nikola, who surrendered his armies before they could at least aid the retreating Serbs.

Milovan Djilas was seven years old when Montenegro was unified, in 1918, with the Serbs and Croats and Slovenes (the latter two peoples having been under Austro-Hungarian rule) in the kingdom that was later called Yugoslavia. He was eighteen when a Montenegrin delegate killed the Croat leader Radić in the Yugoslav parliament and thereby hastened the events that led King Alexander to declare a dictatorship. At eighteen, too, he was already a communist as he went to Belgrade to study at the University—at the close of this volume of his autobiography.

He was twenty-two, in 1933, when he was imprisoned for three years by the royal government for his communist activities. At the prison at Sremska Mitrovica he met a fellow prisoner, Alexander Ranković, who years later, as head of the secret police of Communist Yugoslavia, was to order the several arrests of Milovan Djilas that returned him to the same prison cell, only now for a different heresy. In *Land Without Justice* Djilas asks: "Are men doomed to become the slaves of the times in which they live, even when, after irrepressible and tireless effort, they have climbed so high as to become the masters of the times?"

Even were it not for such historical ironies, and its undoubted historical importance, *Land Without Justice* would be a rare book—even if Djilas were not already a symbol for our times. This is the work of a poet, telling sad brave tales. It could not be otherwise than sad. Extraordinary numbers of Djilas's family, friends, and schoolmates were destroyed by war and revolution. Each had his loyalties, as honest men must, and it was their loyalties that often tore father from son, brother from brother. And who is to tell the right of it, in a land where history makes men choose, if they will not inherit, their faiths? The past echoes through this book. Kosovo echoes here over a vale of five hundred years when Djilas speaks of a defeat in World War I: "the grandeur of the holocaust at Mojkovac was not in victory, for there was none." *Land Without Justice* could not be otherwise than beautiful, for this is the quality of its style. The melodiousness of the Serbo-Croat language is reflected in this English version, translated by Dr. Michael B. Petrovich of the University of Wisconsin. Djilas has the ear and eye of a poet, and juxtaposed with the many oracular statements in this writing will be found colorful and arresting phrases, as when he tells of a Serbian soldier who "spoke in a drawl and softly, like feathers on a wound." There is in Djilas's thoughts, as well, that mixture of precise observation and generalized reflection one finds in Dostoevski. (*The Possessed* might well include a judgment such as Djilas makes in *Land Without Justice:* "Boričić was essentially a good and noble man but an amateur, and deeply unhappy. He had realized nothing of what he had loved and desired.")

Djilas's book does not easily fit into any literary genre, for it moves at its own pace and carries distinctively its own meanings. Yet in the rhythm of his words and in the recurrence of his themes of suffering and joy, death and life, his writing is akin to the epic poetry Montenegrins sing to the accompaniment of the almost hypnotically mournful music of their national instrument, the *gusle*. It is not, after all, to be wondered that during the first year of his recent imprisonment Milovan Djilas was reported to be working on a biography of Njegoš. One feels that he must still be seeking answers to our times as he explores the mind of that manly genius who achieved not only the freedom of his people but also, through a poetic understanding of a harsh world, the freedom of his soul. Of himself Djilas says in these pages: one could not live without poetry.

WILLIAM JOVANOVICH

New York City
February 6, 1958

*Note on the Spelling and Pronunciation of
Serbo-Croat Words and Names*

s = s as in sink
š = sh as in shift
c = ts as in mats
č = ch as in charge
ć = similar to, but lighter than, č—as in arch
ž = j as in French *jour*
z = z as in zodiac
j = y as in yell
nj = n as in neutral
g = g as in go
dj = g as in George
lj = li as in million

PART ONE
Blood and Cannon

The story of a family can also portray the soul of a land. This is especially so in Montenegro, where the people are divided into clans and tribes to which each family is indissolubly bound. The life of the family reflects the life of the broader community of kin, and through it of the entire land.

The story of any Montenegrin family is made up of traditions about the lives of ancestors who distinguished themselves in some special way, most frequently through heroism. These traditions, spiritually so close to one another, reach back into the remote past, to the legendary founders of clan and tribe. And since there are no unheroic tribes and clans, particularly in the eyes of their members, there is no family without its renowned heroes and leaders. The fame of such men spreads beyond the clan and tribe, and through them the soul of the land speaks out. Because of this, the story of a family may be the portrait in miniature of a land.

This was not quite the case with my family. In its past, back to times beyond recall, there had been neither leaders nor heroes. Its story could hardly be distinctive, even for the district of Nikšić, in which lay its kernel from times immemorial, that is, since the legendary Niša founded the city and tribe which bear his name. From his stock had branched the tribes of Župa, Rovči, and others. The Vojnovići, from whom my family sprang, even though a large clan in Župa, were less distinguished than the rest because they lacked governors and heroes. True, they, too, were rebellious, yet also submissive, rayah—non-Moslem subjects of the Ottoman Turkish sultan—down to the middle of the nineteenth century, when Župa was wrested from Turkish rule.

In a land that prized heroism and leadership above all things, to be without either was vexing and shameful. It was like poverty, or even like a sin to which others were not subject.

Sometime in the middle of the nineteenth century, the smaller Djilas clan became distinct from the broader community of Vojnovići, by virtue of its surname, its way of life, and its violent and irrepressible bent. Yet, until quite recent times, the ties of blood and kinship continued with the Vojnovići themselves, with the rest of the district, and even with the separate tribe of Rovčani, though the only bond among them all was the fable of a common founder.

But it is precisely such fables that make up that real existence in which one lives and thinks. Time washes and wrings from tales all that is unimportant, all the daily humdrum, and leaves only the essential, that which gives body, expression, and meaning to life over the generations. This is how human life begins. One dwells with this from the cradle to the grave. By our existence we forge the base of some new truth in which our own existence will be barely noticeable.

Every clan was forced, lest it fall behind the rest, to weave a legend about its origins, a legend it would then firmly believe. This is how the Vojnovići came to believe they were descended from Duke Vojin of Kosovo, the very one who figures in the folk epic about the wedding of Tsar Dušan.* The fable was so recent that they had not had sufficient time to live the part. To refute it was the

* Tsar Stefan Dušan (1308-55) was the great conqueror of the Nemanja Dynasty, who in 1346 extended his Serbian kingdom into a Levantine empire. The Serb power was broken by the Turks at the Battle of Kosovo, 1389.

ancient, and thus unassailable, legend that they were but one clan of the Župa tribe and had no connection with the Duke of Kosovo.

My own smaller clan's fable about its separate origin and surname arose in a similar way. Every such fabrication is based on some truth, on facts so reasonable and easily comprehended that even their fabricator comes to believe in them. But there were tellers of the truth in other clans, and even renegades in my own clan, who maliciously remembered the less pleasant truth.

The legend purported to tell how a certain ancestor was a jumper, who once beat others by a tremendous leap—*djilasnuo*, that is, he leapt. This was the nickname attached to him, and it gave the surname of Djilas to the smaller clan. The region of Nikšić differs only slightly in speech and in custom from Hercegovina, and the word exists in that region, also. Even the lexicon of Vuk Karadžić * has it. Many surnames in Hercegovina, frequently quite nasty ones, stem from nicknames. Our legend arose out of something that was, on the surface, real and reasonable.

The reality was, as usual, rather different.

There is a story told in the Vojnović clan about a frisky widow who once made rather unrestrained advances upon a miller. She leapt at him—*djilasnula*. And this, or something of the sort, led to the nickname Djisna—the Leaper. No one in my family, not even my father, who was more active than the rest in spreading the other legend, ever denied the existence of old woman Djisna, for one was bound to be proud of her sturdiness and determination. Still, everyone evaded telling how she got this name, from which our own so obviously derived. The story about the miller need not have been true; old woman Djisna could have leapt in another manner.

It is painful to destroy legends, but one must.

* Vuk Stefanović Karadžić (1787–1864), a Serbian philologist whose dictionary and grammar established the Serbian vernacular as a literary language.

This irrepressible widow was named Djisna and her sons were named after her. At first they were angry, and blood was shed over that nickname, but nonetheless it held, as in the case of so many others. My grandfather Aleksa finally accepted the surname as his own. He could afford to do so because he was so renowned that his heroism denied any shame. So it was in other clans. They would divide through some more noted and extraordinary ancestor who was capable of imposing a new surname, even when it denoted something shameful.

My forebears were drummed into my head from earliest childhood, as was the case with all my countrymen. I can recite ten generations without knowing anything in particular about them. In that long line, I am but a link, inserted only that I might form another to preserve the continuity of the family, the people, and the human race. Otherwise the earth would be an unpeopled desert with none to tell of it. Man achieves permanence only through those whom he has made to live after him.

Thus I, like so many others, emerged from an ancient tribe of peasants and shepherds. The tribe and clan live through tales of primeval self-awareness. I, too, grew in an indissoluble spiritual bond with them.

My birth took place on an unsettled afternoon at the beginning of spring in the year 1911. That day could have had a special meaning only for my mother. She had to give birth in hiding, for it was considered shameful to give birth within the walls of an unfinished house, and our old one had already been abandoned. For my father the day was important only because, much later, he invented the story that on the eve of that day he had had a dream that foretold that the newborn babe might be more important for the family than were the other children. Maybe he did dream this. Peasants have a way of seeing visions during important events. Just before a war bloody banners are seen streaming across the sky, troubled waters swallow up people and presage pestilence. That day was for me, as for others, unreal. Life began only with awareness of myself, with a remembrance of it, with the formless feeling that I was a part of the surrounding world, that I lived in all its creations and they in me, awakened by the life of remote ancestors.

This was an elusive game of chance. Matter comes to life by becoming to a degree aware of itself and of the world. A speck of dust is suddenly permeated by its own thought, alone transmitting permanence from its own time into the future. The meaning of every individual life, and even of these written lines, would gain were that transmission truer, and thereby fuller and more palpable for those to come. The unbroken existence of man on this miniscule planet, in the fearful infinity of the cosmos, can find glory only in the character and might of the human spirit.

Though the life of my family is not completely typical of my homeland, Montenegro, it is typical in one respect: the men of several generations have died at the hands of Montenegrins, men of the same faith and name. My father's grandfather, my own two grandfathers, my father, and my uncle were killed, as though a dread curse lay upon them. My father and his brother and my brothers were killed even though all of them yearned to die peacefully in their beds beside their wives. Generation after generation, and the bloody chain was not broken. The inherited fear and hatred of feuding clans was mightier than fear and hatred of the enemy, the Turks. It seems to me that I was born with blood on my eyes. My first sight was of blood. My first words were blood and bathed in blood.

Oblivion has fallen on the causes and the details of these deaths, but there remain the evening stories told around the fireplace, bloody and chilling scenes which memory cannot banish. Sparks scatter the ashes from the fire and the embers and flames flare up while words do not let the bloody deeds burn themselves out.

The first tale to take life in my memory was of my grandfather's uncle, the renowned outlaw Marko Djilas. Marko turned against the Turks when the Nikšić district was still under them, and he lived as an outlaw, killing, looting, and burning for twenty-six years. Even now he does not permit oblivion to cover him; a certain cavern is called by his name, and will be, even when nobody remembers the person after whom it was named, until it is

8

rechristened for someone or something more renowned. Marko's exploits as an outlaw reached even to Serbia.

Once, somewhere in the region of Užice, he hired himself out to spend the winter with a well-to-do peasant, and he seemed to be there to stay. But a Turkish beg * moved in on the landlord, so abusing both his servant and his family that Marko's wild nature could not endure it. He got up one night, beat the Turk to death with a club, and then fled back to Montenegro. He was caught and thrown into a Nikšić dungeon to await either impalement on a stake or hanging from a meathook under his ribs. They say that Captain Mušović's lady herself said, "Marko will not leave the dungeon alive as long as that hole is in the ground." Yet, with the help of some Serbs from the town, Marko escaped and sent a message to the lady: "Has the hole been filled by some mischance?" Another time he stole from the Turks a feast that was being prepared. When they searched for the dishes in the hut he sent an empty pot and pan rolling down the hill.

Marko had no progeny. He was married only a single night. He drove his wife away the next morning because, he said, she smelled. A wife and family were not for a man who risked everything to be an outlaw all his life.

There was something too strange about this wild and restless vagabond, who year after year lived in the mountains and in his cavern, playing the *gusle*, the traditional Montenegrin single-stringed instrument, and singing to himself in the lonely nights. To the enserfed peasants he seemed mad, and they gave him the nickname, Mahniti, Marko Berserker. His lone and irreconcilable struggle against the pitiless rule of the begs was premature and seemed senseless. No wonder there were many family legends about him; he had apparently become a legend even during his lifetime.

It is fairly certain that for a while Marko was in the

* A beg (also, aga) was a provincial governor or local chieftain in the Ottoman Empire.

service of Prince-Bishop Njegoš.* There he remained three
or four years, and might have settled down had not the
Prince-Bishop dispatched him, unfortunately, to escort two
Turks who had come to Cetinje on official business. He
found himself with the ancient foes of his faith and nation,
with the brethren of those who had driven him into the
dire paths of outlawry. He was caught between his obligation to his Prince-Bishop and a primeval urge not to let
his enemies slip through his fingers but to find solace for
his heart. The dimly conceived conscience of the public
servant was crowded aside. He cut down both Turks by a
spring and robbed them. Then he took to the hills again.
The Turks had already placed a price on his head. This
time he fled also from the wrath and vengeance of the
Prince-Bishop. He could no longer go back to his unremit-
ting and just sovereign, who had begun to establish order
in the land, and who forgave none who trampled on his
word.

Bishop Njegoš himself, more than anyone else, blazed
and smoldered with hatred for the Turks. He demanded,
not only war against the Turks, but order and obedience in
his state, and he crushed the willfulness of the clans and the
people.

Marko's hard and stubborn nature could never understand that order was necessary even in killing Turks, and
that this was at the insistence of a ruler through whose
windows one could always see Turkish heads drying on
stakes. It was the Prince-Bishop who had gone up into the
plain of Cetinje to meet the *harambashas* † who had killed

---

* Njegoš was Prince-Bishop (in Serbian, *vladika*) of Montenegro from
1830 to 1851. Of the Petrović family, he ruled under the title Peter II,
although he is also known as Njegoš (after his birthplace Njegoši) and
Vladika Rade. He fostered unity and culture from the capital in Cetinje,
and as a prince, a priest, and a poet is recognized as one of the most
versatile geniuses in Slavic history.

† *Harambasha*, a Turkish term, no longer current, here designates
lieutenant or aide.

the mighty Smaïl-Aga Čengić; he who played with his victim's severed head as with an apple. Certainly the Prince-Bishop was right: without order and discipline the struggle against the Turks could no longer be carried on effectively, on a firm foundation. But what was order and discipline for the Prince-Bishop was for the independent clans and willful Montenegrins a loss of their freedoms. Accustomed to every crime and lawlessness, they resisted him, and sought help even from the neighboring Turks.

The Prince-Bishop was hard on Montenegro, but the country was not easy on him. Not without reason did he often call it a lawless and accursed land.

Till his death Marko yearned sadly for his unhappy and implacable sovereign. He never forgot or forgave that he had had to flee from the Prince-Bishop's judgment because of the blood of Turkish dogs. He composed some malicious verses about the lingering death of his great, long-suffering ruler:

> *When Bishop Njegoš lay to die*
> *And gave his soul to God on high,*
> *Three weeks the rain came pouring down . . .*

These verses lived secretly among a discontented people for over half a century. They may have been the reason why Prince Danilo * decided to rid himself of the outlaw, by treachery if necessary. The harsh and unyielding Prince had inherited the firmness of his uncle the Prince-Bishop, but neither his manhood nor his spirit.

One morning when Marko awakened, his cave was surrounded. He was lured out by a pledge of truce and met a volley of rifles. The attackers were led by the famous hero and new district captain of the mighty Ćorović clan, Akica Ćorović. Dying, Marko moved his lips to speak—to curse

---

* Prince Danilo II succeeded his uncle, Peter II (who as a bishop was unmarried), and ruled from 1851 to 1860. He secularized the rule and gave over his ecclesiastic function to an archbishop.

the treachery or to leave a message—but Akica rammed a rifle butt into his teeth and stopped his last words.

Every government newly in power acts without consideration or measure, and the Prince's government had so acted. Marko's death was for years discussed as bloodthirsty and inhuman. An outlaw must be brought to reason, but he need not be murdered.

The Djilas family at that time was less than a handful; they lived in just a few houses. Most of the Vojnović clan, from which it had sprung, had migrated to the Sandžak, in Turkey. There was nobody to avenge the dead outlaw. Marko's brother, my great-grandfather Marinko, was a retiring and industrious man. The blood that had been shed might have subsided and been forgotten had not Akica boasted that his cruel deed had been not only official but also an act of personal whim and passion. This has always been possible where authorities are inhuman, and especially so in my country. Then there rose among the Djilas kin a will more savage and indomitable than Akica's, that of Marinko's son Aleksa, my grandfather.

Two, if not three, years had gone by since the death of Marko, whose personality had caused a new name and new clan to blaze up from the ashes of the humble living and peaceful dying of former serfs. It was spring and Aleksa was plowing the field. His father, Marinko, was tending the flock in the mountain. Captain Akica Corović, accompanied by two soldiers, came riding by the field. He called out a greeting to the lad. Aleksa replied with a murky silence, the only fitting tribute to a murderer. Akica shot back, "Dog, why don't you return my greeting? For I could lay you out to dry as I did your uncle!" The lad left his plowing, hurried home to his mother, and tricked her into believing that his father had sent an urgent demand for his rifle to fight attacking wolves. His mother gave him a blunderbuss from the locked chest. Aleksa intercepted Akica, fired a shattering volley into his chest, and then, with a dagger, carved out pieces of his heart.

To put an end to blood feuds, the greatest obstacle to the unification of the country, the rulers of Montenegro, from the time of Prince-Bishop Peter II, punished without hesitation by taking a head for a head. This time the head to be avenged was a select one—that of a captain, a doughty hero and renowned leader of a powerful clan. The Djilasi were upstarts and few. In danger from the authorities and the Ćorović clan, Aleksa had no choice but to move to Turkish Nikšić. This is what everyone did who owed blood. They became fugitives in Turkey, although more frequently fugitives fled from Turkey to Montenegro. Any Montenegrin who could find a fugitive in Turkey could kill or capture him whenever and however he was able.

Aleksa's flight was deemed justifiable according to the conceptions of the time. But not in the eyes of my father. He usually passed over the affair in silence, though it was due to Aleksa that the Djilas name became known and talked about throughout all Montenegro. With his father, my grandfather, the clan feeling still predominated and it was no shame to flee to the Turks because of bad blood, but in my father the dominant force had already come to be national and state consciousness. To him every tie with a foreigner was indecent. He was secretly ashamed, though without reason. His notorious father had gone over to Nikšić, to the Turks, but misfortune had driven him there and in his time there was no shame in this.

The Djilas clan had been broken; the calamity brought them together. Settling on the city's edge, they dealt in sheep, but lived as on a volcano, with rifle and knife ever ready, defending themselves against other Montenegrins.

Aleksa could not even get married properly. His bride-to-be, Novka, of the powerful and respectable house of Radović, was a young widow, who, when someone asked for her hand in marriage, had announced that she would never remarry as long as Aleksa Djilas was single. But her

family would not hear of giving her to a fugitive. Grandfather heard of her words, and he was goaded all the more
by knowing the opinion of her kin. He took his bride by
theft. The kidnaping of brides had already become rare,
except usually with the girl's consent, and Novka's family
took offense, more so than if a male member had been killed.
Grandfather was never reconciled to his unwilling relatives, and later they took part in his murder. Consequently,
Grandmother was cut off from her brothers, and her children from their maternal uncles.

On the eve of the war with Turkey in 1875,* many
fugitives were amnestied, among them my grandfather. He
returned to the homestead in Župa.

But the sovereign's mercy did not spare his head.

An order went out to Aleksa from the war lord Petar
Vukotić, Prince Nikola's † father-in-law, to pick twelve
of his best sheep as tribute. Many war lords among the
Prince's kin relegated special privileges to themselves, acted
arbitrarily in this way, and grew wealthy on the liberated
territories. So the demand was not at all strange, especially
when asked of a man who had been a fugitive and who
owed blood, even though all had been forgiven. Aleksa
Djilas, however, was not accustomed to paying tribute. He
replied bitingly, "I'll give him whatever he can take from
the point of my sword."

Aleksa's own godfather invited him to a celebration,
prepared secretly for his death. There, at his godfather's
board, a guest hit Aleksa on the head with a wooden mallet.
If they had killed him in a manly way, with a gun and
out of doors, there would have been less hatred to remem-

---

* In 1875 Montenegro and Serbia supported rebels of Serb nationality
and Orthodox faith against the Turkish rule in Bosnia and Hercegovina
and then directly attacked the Ottoman forces (Russian joining them in
1877). In 1878 Turkey capitulated and yielded great territorial conces-
sions to the three Slav allies under the Treaty of San Stefano.

† Prince (later King) Nikola succeeded his uncle, Danilo II, in 1860
and ruled until 1918. In 1910 he proclaimed himself king of Montenegro.

ber! But they felled him like an ox. And they threw his body into the middle of the field.

The authorities in Cetinje had directed the murder; for them not even spiritual kinship was sacred. Many others were tricked in this same manner. Prince-Bishop Njegoš had frequently broken his word, though never willingly, but he, at least, had never forced Montenegrins to trample on their most sacred customs. Prince Danilo did not balk at this, and Prince Nikola dispatched his opponents even more silently and without notice. It could not always be so.

In the Montenegro of that time it was not unusual for whole families to be wiped out, down to the last seed. Thus it was decided to destroy the rebellious house of Aleksa Djilas. The murderers of Aleksa set out to kill off all the males in his family. They surrounded his house and called out Aleksa's younger brother Veljko, who was brave and fast with a gun, and therefore they feared him. Veljko, unsuspecting, came out and was met with a volley of rifle shots. Though wounded, he slipped away in the dark through the bullets and the knives. Aleksa's oldest son, Mirko, a lad of twelve, fled through the window. The middle son, Lazar, lay hidden by his mother in the manger hay. Aleksa's father, Marinko, bent and deaf from old age, was innocently warming himself by the fireplace when the murderers broke in and killed him by the hearth. His blood fed the flames and his body was burned. My father, then a year and a half old, was in his cradle. As a murderer swung his knife, one of my grandmother's kin, who was among the attackers, caught his arm. "It would be a sin—a babe in the cradle!" And so my father lived. No one touched Stanojka, the oldest child, who was fifteen and had just come into maidenhood; it was not the custom of Montenegrins to take up arms against women.

The house and the cattle were plundered. The family was left on the bare bloody rock.

Aleksa's head had to be rescued, for according to beliefs

of that time, a retrieved and preserved head was like the retrieving of one's honor and pride, almost as though a man had not been slain. None dared except Aleksa's daughter Stanojka to go and bring the head, to keep it at least from being gnawed by the dogs or dishonored by enemies.

The night after the massacre was dark and rainy. Stanojka went out to seek her father's head. She learned through the telling of others where it lay, and started back with her treasure to a home left desolate. There was a long way to go, two or three hours by day, longer on a blind night. The Gračanica River had overflowed, the fords were lost from sight, and the swirling waters almost swept away both the head and the girl who bore it into the night.

Stanojka lived long. But always, her whole life through, her head trembled markedly. Though well built and sturdy of frame, she remained somewhat lost from that night on, both in her speech and in her mind. Her brothers and mother, as well as her own children, later, and everyone else knew of the shock she had suffered. And she, too, knew. But she endured her misfortune calmly and patiently, as though she had explained it all to herself: It had to be so; she did what a woman was obliged to do for her kin. She was neither proud for retrieving her father's head from the enemy and from the water that night nor ashamed for becoming dazed and unsettled the rest of her life. She was at peace with herself and with life, seeking nothing, clinging to her brothers and their children, and to her only son. Her brothers took poor care of her, though she was impoverished and without a roof or a foot of land, not so much because of stinginess, but more because they believed that, unsound and withdrawn as she was, she did not need anything, that she was incapable of wanting. She was as delighted as a child over the smallest gift, guarding it jealously however worthless it might be. Life did not shower her with bounties.

Late at night, when all would be asleep, she would get up, stir the embers of the fire on the hearth, and stare bluntly at the dancing flames. Her sturdy feet, cracked and as broad as hoofs—for she constantly went barefoot—were turned soles to the fire, and the hard, gnarled fingers of her hands were clasped about her knees. Hunched up thus, she could sit for hours, until some sound or the dawn startled her. Sometimes, sitting thus, she would hunch over even more and begin to cry, not in any usual way, but without a sound, without a movement, only the tears streaming from large green eyes down a face cracked by sun and toil and plowed over by misfortune, a rock beaten by wind and rain. When she found herself alone in the mountain where she thought no one could hear her, she would wail aloud.

If others forgot, at least at times, Aleksa's death, she seemed fated to be bound to it in every moment and with all her being. If she took joy in anything, it was in that imperishable remembrance. Truly she had something to remember.

In his day the handsome, dark, and blue-eyed Aleksa had been in all things a man of renown. The killing of Akica Ćorović had promoted him to the company of notable Montenegrins. Though the authorities campaigned against blood revenge as a national vice, the people esteemed it highly, especially when it was heroically done. His memory was mighty for his clan, all the more so because of the terrible retribution that befell the house of Djilas after his murder. The family was ruined along with their property, and the clan scattered. His brother Veljko escaped only to succumb, too; he lost his life a few years later. His children wandered off, abandoned, to carry a burden of hatred and bitterness and an incurable ache the rest of their lives. Whoever remained to carry on Aleksa's name met with ill will; the ill will of men of the same blood, faith, and tongue.

It was all according to some strange chance. It seems that families, even more than peoples and states, rise and fall with certain men, as though all their strength had gathered in them. Years and years of travail and misfortune go by, and whole generations are crushed, until the family rises again through a new personality with fresh strength and greatness.

Aleksa Djilas's brutal death horrified all of Montenegro at the time, though it was neither the only such death nor even the most revolting. Every Djilas observed long after how men of other clans would look upon them with sympathetic regard, not daring to lend support, yet ashamed not to pity. The Djilasi lived as though excommunicated, like men marked by some indelible disgrace which could be neither redeemed nor forgotten, all the more so because there could be no feud to wipe it out. To lift a hand against a member of the ruling Petrović dynasty meant not only to forfeit one's own head but also to bring destruction upon the whole clan and misfortune to the state and its people. In their superstitious submission the Montenegrins held the house of Petrović to be sacred. They hardly considered revenge, even in their innermost thoughts.

The reigns of Prince Danilo and Vojvoda Mirko * were distinguished by unbounded cruelty; the long reign of Prince Nikola, however, was marked more by slyness and calculated generosity. Prince Nikola knew, better than anyone else, the good and bad points, not only of Montenegrins as a whole, but of every clan and of every Montenegrin of any note. His was the rule of a wily and tough

* Vojvoda signifies chieftain, or noble. Vojvoda Mirko, a man of savage temperament and high courage, was Prince Danilo's brother. He was passed over as prince, by his own assent, not only by his brother but by his son, Prince Nikola. His strong will and fearlessness fitted him for military command—he led the Montenegrins in numerous campaigns from 1851 to 1878—but not, as he himself recognized, for civil administration.

patriarch in a disobedient and motley family. Prince-
Bishop Njegoš and Prince Danilo never thought of rectify-
ing their errors and injustices; they were busy with the
immediate tasks of order and power. Prince Nikola could
afford to be different, to smooth over offenses, not only his
own and those of his relatives, but even those left behind
by his predecessors. He knew well, and always bore in
mind, that though Montenegrins cannot forgive or forget
injuries and injustices, they are always ready to succumb
to benefaction and generosity, even from the most desperate criminal.

When candidates were picked for the first cadet school
many years after the death of Aleksa Djilas, the Prince
ordered that one of Aleksa's sons should be chosen. This
could only be my father, Nikola. He alone of the children had completed elementary school, and he was brighter
and abler than his brothers. His oldest brother, Mirko,
apart from being illiterate, had had a withered right arm
from the time when Grandfather carelessly wrenched it
while lifting him to his shoulder. Lazar was also illiterate
and generally incapable.

So it fell to my father's lot to go to the school that
trained the first Montenegrin officers. These officers later
replaced the leaders of the so-called People's Army, who
had gained their positions either through personal distinction or, more frequently, through inheritance, and, in any
case, without any schooling.

But Father did not complete his education in peace. He
had distinguished himself in school on the track and in
high jumping. In a contest held before the Prince on
Cetinje's Obilić Field, he leapt over a stallion. The Prince
admired the jump and ordered him to repeat it. While
Father was getting up speed, the Prince signaled for the
horse to be turned lengthwise. Still running, Father was
taken by surprise. He crumpled to the ground, and blood

gushed from his nose and ears. They carried him away half dead. Later he recovered, but though the Prince gave him a gift to make matters right, this only added insult to injury for the embittered lad.

His father's death constantly oppressed Nikola, both the pain and the unredeemed shame. The yearning for vengeance grew stronger and more unbearable in Cetinje, where he found himself near those he knew had dipped their fingers in his father's blood. Within him slowly ripened the thought, the intention, the purpose, to bring peace to himself through the satisfaction of revenge.

Only in his old age did Nikola hint that it had been his intention to kill a Vukotić, one of the Prince's kin by marriage. Even then he would add that he could never have raised a hand against the sacred house of Petrović, and especially not against someone of princely blood. But he really was not telling the truth, even though he might have believed it. It was as though he were trying to convince himself that he had never contemplated taking revenge.

Nobody ever discovered how suspicion fell on Father. One day he was charged with the intent of wreaking vengeance on one of the Prince's sons. He was fettered and thrown into a dungeon. His brothers, whom he had sent to Bulgaria, were brought back under guard and jailed. But they were not detained long. Father remained shackled in heavy irons for a year and a day. Years later the scars above his ankles were still visible. The damp had sucked away his strength and color. But his nimble wits were not lost.

During the investigation my father's head hung in the balance. The Montenegrin court had little jurisdiction, but when a defendant was brought before it, it had respect for the truth. Twice he was taken to the palace at night for talks with the Prince, and on these occasions his fetters were removed.

What did the Prince ask him? What did Father admit?

Father told only this: The Prince took an oath that he knew nothing of Aleksa's murder—otherwise, the Prince swore, may I eat my roasted children on Christmas. Father believed in the Prince's innocence, or perhaps simply wished to believe in it. The Prince's oath may have been a good way for my father to justify to himself and before others why he had foregone revenge forever—that most terrible and most sweet-tasting of passions, which still in those days stirred the breast of every Montenegrin.

Prince Nikola was not bloodthirsty. He preferred to break men rather than kill them. It appears that my father did confess to something—either his desire or his bitterness—and the Prince gave him his life in exchange for a solemn oath of loyalty. Father promised the Prince to forget the thought of revenge. My older brother and I deeply regretted this reconciliation.

When Father was released from jail, his rank was restored, and the Prince even gave him a gift, as befitted a sovereign's grace. He granted lands both to him and his brothers. The lands were on the very border of Turkey, in the plains of Kolašin, in the village of Podbišće, where there was still bloodshed. Father was also made a border officer. The family's material position thus had improved. In fact, however, this was only voluntary and paid exile.

Father's brothers, who were already married, built sod houses on the new land, once the estate of a beg, and began a new life. But in the village itself, there was not enough land for Father. The village land had been distributed already among the new settlers who, after the War of 1875–1877, had descended from the mountain slopes over the Moslem holdings like hungry wolves on a sheep pen. Father was given a parcel outside the village, in an isolated spot, virgin soil in the middle of a forest. It was a little less than an hour's walk over a muddy path through the woods to the outlying houses of the village.

The land had to be cleared, and a house had to be built.

Father at first erected a sod house, just a place to lay his head. For years, frugal as he was, Father saved, and, when he had something, built a solid house of stone, two stories, like the houses in his old home. That house stood, and stands, in the middle of the Podovo bluff, a very windy place, but with a commanding view.

Father's main concern was that his chimney be seen from everywhere around, from all the roads and hills. But Mother, who worked around the sheep and the house, had trouble with the wind and cared little enough for the view. She always cursed Father for building the house on just that spot. Though she was right, we children were truly glad to be able to see the house from afar, on a hill-top, no matter where we were coming from. Sturdy and gray in the middle of the bluff, it greeted us from a distance. There it lay on the green meadow between the Tara and Štitara Rivers, among the crags and hills, waiting to give shelter to tired and frozen travelers.

It seems impossible in life to have something both useful and beautiful. So men are divided. Some are for the useful, some for the beautiful. I placed myself on the side of beauty.

Nearly all the land in the Kolašin region had been taken from the Moslems, whom the Montenegrins massacred or drove away after conquering them. Even their graveyards had been leveled and plowed over beyond recognition. The blood enmity between the two faiths had been so great that the Moslems themselves moved away, abandoning their houses and farms. It was not only the agas * and begs who fled, but also the Moslem peasants, who owned most of the land. The seizure of Moslem lands was regarded as a reward for the horrors and carnage and heroism of war.

And so the neighboring village of Bjelojevići, in the lowlands along the slope of Mount Bjelašica, was settled by a family of the Vasojević clan, from the Lijeva River. To Podbišće itself came families from the Morača and Rovči. The settlement of both villages went with the decimation and expulsion of the old Moslem settlers, who are called Turks to this very day, though they are of the same blood and tongue as the Montenegrins. The War of 1875–1877 had long been over when the Montenegrins surrounded twelve Moslems of the Koljići clan from Bjelojevići, who were fishing for trout in Biograd Lake in a dense forest, and slaughtered every mother's son. It was then that the women and children of the Koljići abandoned their vil-

* Aga (also, beg) is a Turkish title of honor, sometimes used for a general or superior military officer, or governor, in a local district.

lage and migrated to Turkey, while the Montenegrins seized their houses and lands.

Settled by people from the hill tribes, the whole region differed both from neighboring Hercegovina and Old Montenegro. The speech of the highlanders is soft and supple, like that of the Hercegovinians, but it also differs in many ways. Their songs are warmer and more colorful than those of Old Montenegro, though not as realistic and flinty.

With each war and especially after the Great War of 1875–1877, and after each migration, various regions, clans, and tribes became more and more visibly mixed. The parents may still have maintained tribal and other distinctions, but to the children these were only traditions and tales.

We children lived and grew up among these highlanders, who were themselves beginning to lose their tribal characteristics. Father was spiritually bound to Montenegrin Hercegovina, though all our political traditions and state allegiance was Montenegrin. We did not rightly know, like so many others of many generations, just where we belonged. We were among those who belonged to Montenegro in the widest sense, as a whole. We were Montenegrins who had already been assimilated; we had lost tribal differences.

All this crisscrossing affected my family even more because we had come from a distant region and had no kin. My mother was, in fact, not a Montenegrin at all, though it could be said that her family had become Montenegrin by living among them for so long. Everyone must belong to some flock.

Mother's father, Gavro Radenović, had come from Plav, from the village of Meteh; his people were called Metešani. My father would say in anger that the Radenovići-Metešani were Albanians, but this was not true. The Radenovići were Serbs from time immemorial. They were known patrons of Dečani Monastery even in the days of the

Nemanja kings.* Despite all migrations and massacres, they
flourished; they have maintained their homestead to this
very day. The Radenovići became blood brothers with
surrounding clans, which had become Turkish. Only their
perseverance and heroism, and the protection of the
Šabanagići, renowned begs of Gusinje and Plav, whose
tenants they were, kept them alive. Gavro fled to Podbišće
with his brother after killing some Luković Moslems. He
had settled in Podbišće long before my father. It was there
that my father met my mother's brothers and my mother,
whom he came to marry at the beginning of the century.

Mother's kin differed in everything from the Montenegrins. They tilled the soil better. Their food was tastier
and more varied. They dressed better and gave more importance to cleanliness. This was not only because they
were better off—though they did buy the house and property of Beg Zeko Lalević himself—but because their way
of life was different, more orderly and advanced. They had
fled hither from a region in which the people had lived off
the land for centuries. All the others in our village were,
till yesterday, poor shepherds who, by hook or by crook,
had gained some land, unfertile wasteland at that, to squat
on forever.

In contrast to the Montenegrins, or the highlanders, the
Metešani were a proud people, but unostentatious. They
were loath to take up quarrels over words, but they were
prepared to devote inexhaustible effort and invincible
heroism to any issue involving something real or intolerable. They, more than the villagers, thought and lived in a
world of reality. Such, too, were their dress and speech—
rough and without much color. They had neither the

---

* The Nemanja dynasty in Serbia ruled from the twelfth century
until its defeat by the Turks at Kosovo in 1389. Tsar Dušan Nemanja
represents the dynasty at its height—1308–1355. The Dečani (Orthodox)
Monastery, located near Peć in South Serbia, was a center of Byzantine
art and culture.

gusle nor heroic songs. And their women were different. They lived more at home and were withdrawn. They never scolded, unlike the Montenegrin women, all dangerous shrews who, once they begin to abuse others, can never stop. The Metešani did not beat their wives, or at least they did so only rarely. With them a man did not regard it as shameful to take a woman's place in any task.

Was this because they had lived for so long as tenants of the Turks and under the unsheathed Turkish sword, in a constant struggle to survive, in a place where work and silence had become of great value?

At any rate, they were people of that stamp.

Mother's father, Gavro, was a real patriarch, strict at home, and not much for words. Having killed some marauders who had trampled over his fields and molested the women, he abandoned all his property to the ravages of his neighbors, and settled his family in a new place like their native region—with mountain air, swift and cold rivers, mountain and forest overhead, field and meadow around.

But he found the people strange. They lived, dressed, and spoke differently; their ways were different. Had Gavro been a Montenegrin, he might well have boasted loudly of his heroism—of how he had killed Turks in Turkey itself! But he spoke of this unwillingly, and briefly, without embroidery or exaggeration. If anyone came into his house, Gavro gave him hospitality, but never with ostentation. So he was reputed by the Montenegrins to be uncongenial and a shade this side of stingy. The only truth in this was that he was not extravagant, and he managed this among men who as yet neither knew how nor, in many cases, cared to make money. What they did make they easily squandered.

Mother, too, was different from the Montenegrin women. She was cleaner, more industrious, more domestic. She closely resembled her father and brothers—tall, big-boned,

and fair. Unlike her husband, she was taciturn and unim-
aginative. Only when she boiled over with anger did she
utter a sharp word, and then never a vulgar one. Yet she always made her point. She never quarreled with anyone
in the village or with her in-laws.

My father, on the other hand, was a tireless talker with
a boundless imagination. Talking was his great and inexhaustible delight, and he could not live without imagination. He found it easy to get into a quarrel. It was obvious
that, especially for his environment, he was a man of great
and remarkable intelligence and capability. He was not
acquisitive, one of those who always talk of making money
and never do, nor was he a spendthrift. He spent money
only in moments of great decision or sudden opportunity.
But his ventures all proved to be unrealistic and profitless.
He had irrigation ditches dug, with crushing effort, but
the water would not flow. He built and planted where
seeds would not grow. He would buy a new property only
to sell it all one day in senseless anger and bitterness.

Father was sick with the love he bore his children, especially his first-born son. Though he never beat his children,
he would talk and talk to them, giving advice or cursing in
anger. Mother beat the children whenever they became too
much for her, but without cursing or scolding. Her brief
and wrathless beatings were easier to take than Father's
endless counsels and curses. Mother's love was barely
noticeable. She loved and did things without either offering
or seeking love and gratitude. Her wisdom was simple, un-
obtrusive, but real and somehow instinctively infallible
whenever it appeared.

The roots from which a human creature arises are many
and entangled. And while his growth is unfolding, a man
does not even notice how and whence comes the whole-
ness of his personality. The component parts become lost
in it. He is derived from various strands, but he also forms
himself—rearranging impressions from the outside world,

inherited traits, and accepted habits, and thus himself having an effect on life about him.

But who can comprehend it all?

Man does not leave behind the world he found. Though he arose from it, he himself has changed it, while also becoming changed within it. Man's world is one of becoming.

That very unstable world, however rough, is all the clearer for its genuine reality and beauty. It changes no more.

Were the Tara not so swift and cold a river, its turbulence would make it a perfect boundary, bloody and unsettled, not between two states, but between two worlds locked in a life-and-death struggle of centuries.

Here, by the Tara, three brothers settled in a valley to start a new life.

The border did not go along the Tara, but cut across above Mojkovac, toward our village, and then turned off over Mount Bjelašica. The whole region is hilly and wooded, with slopes on each side. Even had there been good will, it would have been difficult to prevent incursions. But there were grave violations. The looting and killing was ceaseless. Neither side, the Montenegrins or the Moslems, permitted their guns to be silent, as though afraid to forget that their scores remained unsettled since Kosovo. It was a life of deceit, on both sides, of ambushes, sudden attacks by day and night, and the rifle ever loaded, both in the fields and in the mountains.

In the spring, when the hills grew green and the Tara descended, the bloody activity would begin. Meadows, murky at night, and black forests, which stretched from the heights to the water's edge, where the sounds of the river changed constantly, unexpected shots and death, chases and ambushes—here was my father's service. Several times my father walked into an ambush at night. But the Turks fell into his, also. It was as though both felt the delight of horror, and permitted each other to come quite close.

One night Father went out in front of the house, where a few fallen beech trees lay, and a shot rang out. Father returned fire. But the night has no eyes. Father suspected that it was not a Turk, or else the man would not have crept almost to the very threshold, as if he knew even where the chickens laid their eggs.

These unparalleled years, in which the forests were felled and the children born who were to take the place of the first settlers, live in tales and memories with the freshness of something wild and the horror of a suspected ambush.

Near my village in these years the son of the man who had enticed my grandfather to his fatal dinner lost his own life. The godfather and his brothers had moved to Turkey, near Šahovići, probably to be safe from the Djilasi and to give people time to forget his crime. But in going to or from Montenegro, they had to pass through the village in which my father and his brothers had settled. So suspicion fell on the Djilasi when this man was murdered, though, since he was from Turkey, there was no investigation. This murder by night remained forever dark and mysterious, and tantalized us children, more from a desire to learn whether Grandfather had been avenged than from curiosity or pangs of conscience. Father denied that he or one of his had been the murderer. But he did so with an intentional lack of conviction: No, it was not a Djilas, God forbid! The dead man was not a criminal, nor even the kind of man on whom one could revenge oneself as was fitting.

About this time Uncle Mirko was wounded. He was a border guard, and, in chasing a smuggler, ran after him into Turkish territory. The fellow lay in wait and wounded him badly in the chest. That wound, which my uncle carried unhealed for several years, was to be the cause of his death.

Years and years passed after that event, but my older brother and I, when passing through the clearing beside

the place where he had been wounded, never forgot to
stop, silent and choked, by the rock on which Uncle
leaned as he lost consciousness. The people believe that
no grass grows on the spot where human blood has been
shed. On the place where Uncle fell, the grass was dark
red, even in spring, as though it had drunk of human
blood, our blood, Djilas blood, from which we, too, had
sprung and which coursed through us.

I felt this emotion with particular force whenever I
passed by alone. It would begin to seize me even before
I reached the spot. And, as I came to the rock, it seemed
that everything alive in the woods had frozen, mute and
motionless with the pain of sensing a human death. Not a
sound, except the pounding of the heart. And I saw nothing
except a little bug silently milling about the rock on which
Uncle had fallen, thus deepening the stillness. I could
neither stand by that rock nor tear myself away from it.
It seemed made for a wounded man to lean against. It
was as though the wound had just been opened, with the
spilled blood steaming and spreading.

Everywhere on the roads wherever we went, there
was sorrow—tombstones and graves, murder and misfortune, one after another. The murder of enemies was forgotten, but our own Montenegrin losses, especially if
caused by a brother's hand, remained fresh in the memory.
One no sooner passed a mound and put it out of mind
than another waited around the bend. Every stopping
place had a grave.

And Uncle was not even avenged.

Uncle Mirko was generally an unlucky man, and all of
his misfortunes were incurable because he had no male
heirs. He was older than my father by ten years. They
loved one another dearly, with the kind of love a father
and son have for each other. He had three married daughters, whom he could not endure, most probably because he
had no sons. He never worked much, nor could he—with

his unhealed withered arm. Father helped him with money. But only him. His property was quite neglected. He had no will to care for it. Why should he? For whom? He was fated to be a lone breadwinner among his fellow Montenegrins, as though he were cursed and proscribed above all men.

Mirko was a strange mixture of courage and caprice, beauty and ugliness. Handsomely built, sturdy, but not short, he had big moss-green eyes, a forehead hewn of rock, and wide black mustaches. He liked handsome weapons, dress, and a good horse, though he was a pauper. He, too, went to war, but, because of his disability, always on a horse, even during attacks. He never had the fortune to be killed. This is the kind of man he was—ready to do any manly deed, and yet he was not ashamed to mount a horse and scatter gypsies who had raised their huts on used pastures. His word was good, and he was hospitable. He knew how to be tender and humane, but also cruel and thoughtless.

My other uncle, Lazar, was an egoist, dull-witted, withdrawn, and inept. In war he was a poor soldier. It must be said, however, that he never boasted of heroism. He married twice, but he had no children by his second wife. He beat both his wife and children, in fits of fury, which took even him by surprise. The next day he feared the wife he had beaten yesterday: Would she be good and tender? His sons did not remain in his debt. When they grew up, they beat him.

He pretended he was deafer than he was, whenever he found it convenient. Yes, he liked to pretend in everything. Short, plump, slow, and well built, he liked to quarrel, though he avoided fights. Once in a fight, he would cast aside his rifle and pick up a club, so as not to kill anyone lest blood come to his eyes. He was loutish and somewhat eccentric, without much sense, sly in a primitive, and at times amusing, way, but—sly. Nobody had much of a

liking for him. But neither was he very lovable toward
anyone.

After he had migrated to the region of Kolašin, Uncle
Lazar went to work in America and spent three years there.
He learned hardly a word of English, and earned no money.
As he went, so he returned. But bad luck followed him
there: his wife had thrown off her yoke and had become
pregnant by another man. To save the honor of the clan,
my father almost had to compel him to drive her away.
His sons grew up undisciplined and on their own, largely
in my father's house, only to wander throughout the
world as soon as they broke away.

Lazar was, of course, dogged even in war by mischance,
but of the kind funny stories are made of. A friend of my
father's made him an assistant, actually more of an orderly.
During a lull in some fighting a stray bullet struck Lazar
in the back. On the operating table the doctor extracted
the bullet with pincers. The bullet had passed through his
pack and, already spent, barely punctured the skin. But
now he, too, had suffered a wound, though he did not like
to boast about it. Another time, he fled from a battle down
along a ravine and, in his haste, stumbled upon a wounded
man. The latter implored him to save him, and so Lazar
dragged him away, leaving behind a wide track as though
a harrow had gone by. Later he asked for a medal—for saving a wounded man under heavy fire.

My father shared something with both his brothers. He
was garrulous like the older brother, and quarrelsome like
the younger. But he was more garrulous than the former,
and less quarrelsome than the latter. In everything else he
was different.

He belonged to that first generation of Montenegrin
officers who had any sort of education. But educated or
not, they remained peasants in their way of life, in their
speech and behavior. They all lived in villages, in houses
that were somewhat more handsome; they dressed in fan-

cier clothes, and ate better, but they, too, busied themselves with their cattle and land like all the peasants, hiring help only for the heaviest tasks—plowing, digging, and mowing. All these half-educated officers, teachers, priests were easily identifiable in Montenegrin hamlets before the last war. On market days they talked about politics in the coffeehouses, sipped brandy and disputed endlessly about Russia and England, the Croats and Belgrade. They still voted as tribes, always discontented, and dressed half in national costumes and half in city clothes, a new shirt under their gold-covered tunics, or a coat.

It was these men who bore the brunt of the wars of 1912 and 1914,* fulfilling their duty bravely and honestly. They replaced the generation of officers who had fought the battles of 1875, and were substantially different from them. Their personalities were less developed, for they had not risen through personal bravery and in rebellions against the Turks. But if one of them happened to be a weakling or a tenderfoot, he did not count for anything. They were foreordained to be steadfast and devoted servants of the Prince, picked men all, to be sure, but without the pride or the backbone of the more distinguished representatives of preceding generations. Hardly one of them even became a rebel against the Anointed One of Cetinje,† whereas the most renowned war lords of earlier generations,

* In 1912 Serbia, Montenegro, Greece, and Bulgaria opposed and defeated Turkey in the First Balkan War. Friction over the division of Macedonia, which had for centuries uneasily harbored Serbian (also Montenegrin), Greek, and Bulgarian inhabitants, led Bulgaria to attack Serbia in 1913. The Second Balkan War began and ended that year as Greece, Rumania, Turkey, and Serbia quickly defeated Bulgaria. Austro-Hungary then became apprehensive of the growing influence of Serbia in her Serbo-Croat provinces of Bosnia and Hercegovina. On June 28, 1914, Archduke Francis Ferdinand of Austro-Hungary was assassinated in Sarajevo (Bosnia) by a Serb patriot, whose shots were the first of World War I.

† Meaning Prince Nikola; Cetinje was the capital of Montenegro (the capital is now Titograd, formerly called Podgorica).

such as Marko Miljanov, Jole Piletić, and Peko Pavlović,
soon after the Great War of 1875 came into conflict with
their arbitrary and grasping sovereign. They left their
homeland and scattered their bones in foreign soil. Only
later did a younger generation of men, all educated abroad,
in Serbia, rise to rebel in another way against the corrupt
camarilla and already decadent Prince and King Nikola.

Who is there to say from which of these strands I
sprung? It was from an environment of peasant civil servants, more peasant than anything else, like so many Montenegrin intellectuals of my generation.

Our household was completely peasant, but it lived a
better, a more civilized life—if one can use that term—than
the average peasant family. There was always coffee and
brandy in the house for Father and guests; there were un-
matched plates, but nevertheless plates, and coverlets of
down and blankets, and even comforters. There were always fleas, and frequently even lice, though Mother waged
ceaseless war on them. In winter the cattle were kept in
a manger on the first floor, and during warmer days the
laden air of the manger was overpowering. I did not wear
any underpants until I entered high school, not because
of poverty, but because of custom and our way of life. It
was long until I could get used to their slippery softness.

Though we engaged hired hands and sharecroppers, Father himself always worked on the land, and Mother la-
bored like other peasant women—even more, for the
increased demands of her educated children all fell on her
shoulders. She was illiterate to her sixtieth year, until the
last war, when the death and revolutionary activity of
her children impelled her to learn what was becoming
of her family and her country.

During my father's border service the so-called Kolašin
affair erupted. Some officers were condemned for plotting
the violent overthrow of the Prince's absolutism. My father, too, took part in the arrests and the search of the

houses of the accused, but he pretended not to see one bomb he found. There were executions and horrible deaths in the Kolašin affair. But the most horrible memory was the whipping of the arrested men with wet ropes. People condemned this even more than the executions. Until that time no one had ever beaten Montenegrins during an investigation, except for robbers in their own country and then only by their own authorities. Their human dignity had never been affronted by beatings. They have now become accustomed to others trampling on their human pride, but they have not changed their opinion about those who do so.

The War of 1912 found Father on the border, charged with the duty of inciting border clashes that might serve as an excuse for war. One dawn he led his villagers in an attack at Pržišta, which dominated the approaches to Mojkovac. The struggle with the border guards was a very bloody one. My father leaned a ladder against the sentry house, climbed up, and threw a flaming shirt on the roof. Some forty sentries were shot to death or burned, without a man giving himself up alive. In their enthusiasm, our side had underestimated the bravery and resistance of the Turks. There were many dead and wounded on our side, too—all young men, who died eager for war and blood and greedy for glory. There were among them children of fourteen, who had run to battle while their elders watched.

The greatest heroism was shown by the Moslem Huso Mehotin, a renowned freebooter against the Montenegrins. From his earliest youth he had lived at sword's point with them. When he heard shots in Pržišta from his village, he set out against the Montenegrin army all by himself. The last wisps of smoke were rising from the sentry house when the Montenegrins heard Huso's challenge from the neighboring hill. He called out to the Montenegrin officers, announcing his coming, and dared them to wait for him if

they were mothers' sons. Some ten soldiers lay in wait
for him as he strode forth into battle without hesitation, as
though he knew this was the last time his empire and his
faith were to fight on this soil. He fell riddled with bullets
and unsated in hatred, and with him ended the marauding
bands from across the Tara. Only the stories and legends
remain, and they slowly sink under the weight of new and
more significant events. And so Huso Mehotin and his last
act will be lost.

In the year 1941 my brother Aleksa burned down an
Italian *carabinieri* sentry post near this place, at Mojkovac,
in the same way that Father burned down the Turkish one
at Pržišta. My brother told me that it was Father's deed
that gave him the idea. And so it goes from generation to
generation.

The Montenegrins advanced into the Sandžak, meeting
barely any resistance, and that largely from the Moslem
inhabitants. Heroism and glory were easy to come by,
until that insane massacre at Scutari, which enshrouded
Montenegro in black.*

In one battle Father set out after a certain Turk. Both
sides stopped to see what would happen. The Turk ran,
turning to shoot at my father, but he had no time, because
Father was waving a sword just behind his neck. The
sword caught up with the fugitive, and the head rolled on
the meadow. In the telling, this turned out better as a picture than as a recital of great heroism: a head cut off, with
a spurt of blood, while the victim was on his feet and still
running. Somehow no one noticed, either during the telling of the story or after, that this was a human head.

The Montenegrins wiped up the Sandžak in one sweep
and turned on Metohija. In the beginning my father had
been made commandant in Bijelo Polje, but he was soon

---

*Following the defeat of the Turks in the First Balkan War, the
Montenegrins and Albanians attacked each other in the area of Lake
Scutari, which lies between the two countries.

transferred to Djakovica to join the gendarmery flying squads. The family first went with him, then returned home. Conditions were unsettled, and the family was un-accustomed to city life. Life was best and surest on one's own land and in one's own village.

Because of the Balkan War, Turkish carpets and rugs appeared in our house. Had they been bought? Or presented by someone to the representative of the conquering power to ingratiate himself and establish friendship? There were not many of these objects nor were they of great value. But their vivid hues and geometric designs blended in my memory with the peasant Montenegrin songs from my childhood.

So it has always been here: one fights to achieve sacred dreams, and plunders and lays waste along the way—to live in misery, in pain and death, but in one's thoughts to travel far. The naked and hungry mountaineers could not keep from looting their neighbors, while yearning and dying for ancient glories. Here war was survival, a way of life, and death in battle the loveliest dream and highest duty.

First memories are tales come to life of what happened before the remembrance of things. Even so, these memories are no less vivid than of real events.

It must have been in a town, for the floor was yellow and scrubbed and the walls were white. A boy had locked a door from the inside and found himself stuck between the wall and the bed. Through the window, from the stairs, they promise him everything and beg him to open the door. He is no longer a little boy, they tell him, he is two years old. He comprehends. He would like to do this great deed, but he cannot get loose. Someone pushes the bed away with a pole, and he runs to the door, opens it. The light bursts in, and then embraces.

Then they are traveling somewhere, through a dark maze of hills. They ford streams, but do not let him wade. He himself sees that the water is swift and deep and black, though one can still see through it the round white rocks. A soldier seats the boy on his neck, and the boy takes the soldier's cap in his hands. It is new; it might fall and be carried away by the water. The boy cannot understand at all why they are praising him because of this and admiring his intelligence, though he takes delight in their retelling of it. It is as though he could feel by touching that he is already truly intelligent.

Then another town. A very good man lives here—Arif. He is somehow different from the folks at home, for they praise him, even if he is a Turk, even if he is of another faith. Arif's head is bald and long, and the boy is his good friend. Arif's pockets are never empty nor are they guarded by hard and mighty fists. Arif himself says so, because the boy reminds him when he must go to the mosque and goes with him. Arif enters a large building with a high minaret. There are many sandals and slippers on the wooden porch. The boy cannot go into the mosque. He is forbidden, because his faith is different.

Then a warm rain falls, in large hard drops. The boy hides from it under a table. Under the table there are many boots and swords. The boots move as the men talk endlessly. The boy falls asleep and a man who is not Arif carries him in his arms. They say it is his father. There is no more rain or boots. They say the boy was lost. But he cannot understand it—how, why is he lost? Only stupid and bad children get lost, in a forest with wolves. Everyone is frightened: he might have been kidnaped and butchered by Albanians. Why would they steal and butcher children? They are of a different religion. But Arif is an Albanian and he likes children. Perhaps what they say of kidnaping is only a story to make children obedient and keep them from getting lost, because grownups do not like to look for them.

But this is already a clear remembrance. It happened in the spring of 1914, perhaps even before.

On a flattened hillock, one foggy morning, there were officers and their wives, among them also my mother and father and certainly Father's orderly. In honor of something, a cannon was to be fired. Everyone made a terrible face and stopped their ears as though something dangerous was going to fly into them. Father's orderly buried my head between his legs, and I stopped my ears myself. Why did he bury my head in so shameful a spot? I was ashamed because of this but I was afraid to pull away my head. I never told anyone about my shame, nor about the spot where my head had been buried.

The cannon bellowed much louder than a bull, but not intolerably, almost gaily, opening its mouth all the more and bucking as though struck by a whip. About it spread gray-blue smoke. I could not comprehend why people were afraid in advance of that gentle smoke and the firing, which was loud, to be sure, but which did not strike nor cause pain.

Cannon thundered through my first real remembrance. Bombs boomed and rifles cracked throughout my entire childhood, wounding every dream and destroying every picture.

That first roaring of cannon stands framed in my memory, fixed and tangible. Only a woman in a long white dress, tight around the waist, stood warm and radiant by the cannon, the only person among all the rest to stand out clearly in my memory. Neither Father nor my brothers, not even Mother or Grandmother do. On her head the woman wore a broad hat, as if made of the sun. Was it Mother? No, it was not. Mother always wore the national costume, and she was not as radiant and warm. And how was it that this woman remained in my first memory in such bold and fascinating relief, together with the wild roaring of an iron bull?

The gaps in one's memory are enormous.

The war goes on, uninterrupted, only the enemies are different—the Turks, Albanians, Bulgars, Austrians, and Hungarians. We are living in our own village, in a stone house lost amid the wooded hills. The hills are grim and loom up to the stars.

Now there is constantly with us, with me, a woman, warm like the one in the white dress: Aunt Djuka, Lazar's wife. Her throat is round and her bosom is soft. She has no children and loves us, the children of her brother-in-law, as her own. Her love is always the same—warm and tender, without the severity and wrath which flare up so suddenly in Mother. Aunty tells frightening stories about devils,

witches, and demons, all at night, when another, unknown
and terrifying world comes to life. I nestle against her
warm body and try to shut out the fear through sleep,
while behind the door, where the squash is kept, something
grows big, expands and rolls toward me soft and huge, but
does not succeed in crushing me. . . . In my first dream
that night, or one like it, I am scratching a yellow cow
under the chin. She enjoys it and stretches out her red
muzzle. I offer my face to the cow to be licked, but the
cow refuses. I awaken from sadness and pleasure with the
dawn and the stirring. I caress my aunt's throat. Her throat
is smooth and delightful, much more so than the cow's.

But neither the delight of Aunty nor her scary stories
could drive the war away. It came upon us suddenly, its
tumult causing the hills to quake and the heavens to
tremble, drenching every word and every movement in
horror and pain. The grown men had long since left, leaving behind only yearning and anxiety, and songs about
them. Frequently women went to the army, loaded with
food and gifts. Frequently, too, the gifts came back, the
socks, shoes, and leggings; the ones they were meant for
had lost their lives.

Now the tumult of war came quite close, and the shaken
hills were angered. Even old man Milija went off to war.
He was so old that the children would rather play with
him than with anyone else. Toothless, hard of hearing, and
half-blind, he could barely move. But his hands were still
strong and his gnarled fists tough, like oak. It is told that
he himself, in his long lifetime, cleared half of the village
pastures and meadows. He did not have a gun, nor, being
a peaceable poor man, had he ever been to war before. This
time he went, armed with an ax, though war was not like
clearing forests. They laughed at him for this, yet they
admired his heroic age. Old men and boys, rich and poor,
go to war when it comes to their doorstep. And there war
was always just outside the door.

Nothing else has remained of my childhood up to then. It was shattered by the noise of battle and bloody defeats.

For several days the famous Battle of Mojkovac had raged, the last and strongest resistance of the Montenegrin army to the Austro-Hungarian army, there by the Tara, on that same border between our country and Turkey. Mojkovac is about an hour's fast walk from our house. The battle took place on the heights above it.

The Montenegrin artillery was located in part on Mali Prepran, opposite our house. It included a howitzer, whose firing rose above the hum and the roar with a burst that set our frightened windows to trembling. Now it was no longer a matter of hearing news of massacres, of the inconceivable slaughter of human bodies, and of the terrible and comical occurrence of war. Now it was taking place here before us on the surrounding hills, in the woods kept from their slumber beneath the snow by the noise of battle.

The first to appear were the Serbians. The Serbian army was retreating from Peć via Andrijevica, Kolašin, and Podgorica, on the way to Scutari. Our house was four hours downstream from Kolašin, through a wooded and almost uninhabited region. What had diverted the Serbians from their path? They were not deserters fleeing from their army. They could not have wandered away so far. Perhaps these were men unfit for the army, looking for a gap in the front, or waiting for the struggle to end so they could go home. But no, most probably they were worn and wounded troops of the Serbian artillery, sent as reinforcements to the Montenegrins at Mojkovac. They appeared from the deep and bloody darkness of war—and they disappeared into it again.

The news had spread of their dying along the roadside of hunger and fatigue, each dead man with a coin placed in his mouth. But gold does not keep the soul from escaping! There was little bread, and some who had gave none, nor could it be bought at any price. It seems as though our people pitied the Serbians more than they helped them. They did not even help as much as they were able. The Serbians passed through Montenegro as through a foreign land, savage and soulless.

Five or six Serbians stopped at our home. One of them remained forever in our memory. Spare, dour, tall, he wore a black fur cap, and his head leaned crookedly to the right from a wound. He called all the males, even the boys, "brother," and all the women, even Grandmother, "sister." He spoke in a drawl and softly, like feathers on a wound. He ate a big piece of cornbread and a slice of cooked bacon my mother gave him. He thanked us in his supple accents and buried himself in the food so that not a drop of fat from the bacon might fall, no, not a drop, as though his life depended on it. All of the Serbians went, and he, too, went. Yet behind him there remained with us children the soft words brother, sister.

I knew even then that the grim Serbian guarded every drop of fat because of hunger. We still had bacon enough in our attic, and yet we gave him a little piece, smaller than one gives to a child. In war nobody has enough. Everyone is hungry—both those who have and those who have not. The have-nots because they have not, and the others for fear that what they have will disappear.

Once, on a clear day, the clouds scattered, and the winter sun suddenly blazed forth vast and warm, as though we were seeing it for the first time. It browsed happily and lazily in the celestial meadows. The sun was hardly moving when from the hills there shot into the blue two planes —one huge, slow, and black, while the other, lighter and faster, golden and lively as a bee, flew into the sun and lost himself in it only to emerge the more radiant. The wounded did not wish to leave the house, but the women caught up the children and began to run. A little soldier with thick bandages around his head said the house was of stone and large and isolated, so that one might suspect and drop bombs. No bombs fell; the clear of the morning remained untroubled, except for rare bursts which thudded softly in the deep snow.

The planes disappeared at last and the lovely day enticed the children out of doors. Ammunition was spilled everywhere about. Our games were full of peril but wonderful. We would bring a glowing coal on a poker and place bullets on it, then run around the corner of a wall while the bullets exploded merrily. My elder brother led us in that forbidden game. There was nobody to stop us.

And this is my first clear memory of my brother, who was pale, thin, and had big ears and two big buckteeth. He shivered on that clear day, a serious and glum child, even in play. The elders were angry at my brother because of the bullet game—not because we might get terribly hurt, but, rather, because of something else. How could we have fun in such times? We have a presentiment of something terrible and unusual. We hear talk of a bloody battle to come which no one knows who will win. Its meaning is incomprehensible to us. There have been other battles before this. Why should this one be so terrible?

I would not have known that Christmas is a great holiday had it not been for the bloody and great Battle of Mojkovac. This time there was no search for the Yule log, no visiting, only the cannon to greet the unbidden guests.* The cannon's victims were wrapped in blood and pusdrenched bandages. They dragged themselves, one after another, through the snowstorm in a trail of blood and were buried in the snowdrifts. And so all the three days of Christmas were shattered.

The war came with all its force into our house, even

---

* According to Serbian (and Montenegrin) custom, on the day before Christmas (*Badnji dan*) householders go to the forest to find a Yule log called the *badnjak*. After the Yule log is carted to a house, and laid on the fire by means of a highly symbolic ceremony, Christmas morn opens with the householders firing their guns to signal the start of the joyous holiday and to welcome guests.

though it was isolated, for the roads were choked. Everywhere were the wounded, the shells, the guns. The children stopped playing at firing bullets; now the grownups were at more serious play.

There was a ceaseless thunder and flashing on all sides. The cannon never stopped, as though they could feel no fatigue. And whenever a howitzer went off the mountain bellowed from its womb. The thick white beds of snow could not soften that blow. And so it went on without ceasing. The night was beaten and churned, as an ice-covered puddle is by the hoofs of cattle. The maelstrom of battle mixed everything together—men and trees, sky and earth, days and nights. So it goes when men die to defend something dearer than their lives.

Up on the mountain, Mali Prepran, our army walked and walked two whole days, onward and onward, slowly and in disorder, mob after mob. Some soldiers wandered as far as our house. They were angry and sad and prophesied betrayal and catastrophe. What was happening to this undefeated army?

Then came a silence, so complete that not even a bird chirped in the hills. And with it came terror.

Then a morning came whose white silence was once again shattered, but more terribly than ever. There were rumors that our army was retreating toward Kolašin and farther, though it had not lost the battle. Soon the Austrians and the Schwabs would come. Our cannon had to be destroyed to keep them out of enemy hands. A fierce frost and heavy snow could not stifle the shrieks as the iron, which only yesterday tore at sky and earth, at trees and men, now ripped into itself. Blown into a thousand pieces, the muzzles of the cannon roared for the last time, in pain. All the void trembled at the sound, and the firmament shook. The earth must have trembled like this when dragons leapt from mountain to mountain and, falling into the lake, raised the waters to the mountaintops. But dragons

fly at night and appear as flames which set fire to the grass
and wasteland in times of drought. This could not be seen
at all. The wounded air shrieked and the frightened oaks
groaned.

Our folks and the villagers seemed to crowd into a corner. The women held our hands and at night clasped us
firmly to them. Something yet was to happen, more awful
than any terror from those nocturnal tales: the Schwabs
were coming. They were men, just like ours, yet they
spoke a language that could not be understood. Maybe this
made them feared all the more.

Nothing happened. For two days the Austrian army
came through, helmet to helmet, in the footsteps of our
own, but more orderly, even though somewhat slower than
our men. Their weapons gleamed in the winter sun, cold
and sharp. Not all had passed by nightfall, and in the morning they were still advancing—tamping hard all our soil.

In the first Austrian patrol, which came by after several
days, were blond and strangely merry men, who tried to
play with us. We would have none of it, not with the
enemy, though the fear soon left us as we perceived that
they were men like all the rest. Some even spoke our language. They said something to our women. It seems they
were searching for weapons. No one gave them anything,
yet they left content just the same. Long after, people told
stories about them with happy curiosity and wonderment,
how they disturbed no one, how they did not take even a
drop of water.

Begun in that terrible upheaval, my childhood became
no happier and life no tamer when that battle passed away.

A whole people—the Montenegrin—which understood life in terms of war and glory, stopped fighting. A people's army and a state had ceased to be.*

The fall of the Montenegrin state did not blunt the forces of heroism and of manhood, and it seemed to sharpen others —forces of violence, untamed and unrestrained.

With the tragedy of a whole people came also other misfortunes. Or did I begin just then to notice them?

Men became bad, rotten, unwilling to give one another air to breathe. Bestiality and scandal at home, in the village, quickly crowded out of the mind the national tragedy. These vices were our own, Montenegrin and domestic. Though they multiplied with the occupation, the Austrian forces had not brought them and inculcated them into our people. Our people bore this within themselves. It had taken only a relaxation of the reins.

Uncle Lazar was particularly tireless in committing every evil thing against his family. He had a sullen mare which he had brought home from somewhere during the war. He set out to Kolašin on the mare, and, as ill luck would have it,

* In 1914, during the first year of World War I, Serbia and Montenegro successfully opposed Austro-Hungarian forces. But in October 1915, Bulgaria joined the Central Powers and attacked the Serbs and Montenegrins from the east and south, and at the same time Austro-Hungary obtained German reinforcements in the north. By the end of 1915 both Serbia and Montenegro fell, the latter collapsing on the western flank, at Cetinje, when King Nikola ordered surrender. But on the eastern flank the Montenegrins were still holding at Mojkovac at the time of the surrender by King Nikola.

he noticed some soap off the roadside, an ever-enticing article for a villager. He dismounted to retrieve it. While he was making his way to the soap, the mare tore loose and broke into a gallop down the river, over hill and dale. He could not catch her, and returned home angry and frowning, swearing and cursing both the nag and himself and everything else that came to his tongue. His daughter Rosa, who had just entered maidenhood, rebuked him for not watching when he knew what sort the mare was. He seemed hardly able to wait to take out his wrath on someone. He cursed and beat her and his wife. Rosa's weeping enraged him even more. He took her black braids, caught them in the door, and then began to trample on her. My mother rescued the girl, who wept and wept, trying to gather her tumbling hair under her kerchief.

Not long after, Uncle Lazar, embittered, left his land and house and went to Metohija with his wife, his younger daughter, and younger son. Rosa and his older son hid to escape going with him, and they remained with us, though Rosa's feeling of injustice and sorrow did not go away with her father. With Lazar went that bad-blooded little mare, to the joy of everyone.

Uncle Mirko grew even worse, though mostly against himself. He beat his daughters, long married, and drove them through the village. One of them, who took after his evil nature the most, taunted him at the edge of the village to come to her, lift his whiskers, and kiss her shameful parts, and she bawled it out without mincing words. This brought dishonor and censure on the entire village. Mirko drove away his own wife, already an old woman, and she fled to us in the middle of a winter night. Completely decrepit, he nevertheless chased tirelessly through the village after the war widows. His wife claimed that he was a dog and would even attack a snake in the eye. Twilight and dawn found him on horseback, wandering in wild restlessness. In his gay moods, he scattered his money; in his bitter moods, he picked quarrels. Soon Uncle Mirko took off for his homeland in Nikšić, to spend the occupation with relatives, still restless even there.

There is much evil and sorrow in a national tragedy. If Father had not returned soon, it would have seemed as though there was nothing beautiful or tender in life. Only then was I able to get a clear picture of Father, though I long knew about him and had felt his constant presence despite the war and the distance that separated us. He arrived on a high and slender mare, himself slender in his gray uniform, all trim, in boots and with a revolver in his belt, the one he boasted the Austrians did not touch, though they met him along the way. Father was thin and gray and gaunt, like a wolf which runs and runs through the mountains. That slimness and lightness made him handsomer and more tender toward us. He yearned for his children and home with a desire that could never burn itself out.

He had spent the war in Albania, actually in occupied Metohija, constantly fighting rebel Albanians and, toward the end, the Austrians, who had arrived in pursuit of the exhausted Serbian army. In the struggle with the rebels, slyness and skill were needed as much as courage; in battles with the Austrians, which did not actually last long, one needed perseverance, for their artillery pounded away at our troops, rather poorly armed for battle against such an opponent.

From the war Father brought with him, apart from his revolver and horse, several tales—the touching story of a Czech doctor and another and different story about the death of Iso of Boljetini, the Albanian war lord of a village near Kosovo.

The Czech was young, blond, and ruddy, like all Czechs, and overjoyed that he had been wounded and thus captured by brother Slavs. My father was likewise glad and ordered that he be bandaged and looked after. Father continued the attack on the enemy, and when he returned, he found the happy Czech with his brains blown out. Some Montenegrin with a yen for his watch had put a bullet right between his eyes. The murderer was not apprehended, nor did anyone bother to look for him, though it was regarded as a dirty business. This story seemed particularly sad—perhaps because Father drew for us such a vivid picture of the unknown Czech that we children knew him better than we did our neighbors. Perhaps it was also because he had expected to live with his brethren, and had found death at their hands. We long mourned for the Czech and looked with him down the single black maw of that muzzle, unable to believe that it would spit fire. Did he die in a twinkling, not realizing how vainly he had rejoiced over his brethren and that a brotherly hand had killed him? Or had he seen all and known all?

With his flying squads Father was retreating to Podgorica. Disorder reigned; the Austrians had not yet reached the city, where the local administration had already dissolved. What was left of the Montenegrin government, namely General Radomir Vešović, ordered my father to keep his troops intact.

My father knew Vešović well. He was a strange man in whom were combined the traits of insane heroism and extreme vacillation. During 1913 Vešović had been chief of the Montenegrin administration in Metohija, a kind of military governor of that region, known for his too severe administration. He respected and even liked Father, and was always glad to see him, both then and in the postwar years.

Thin, black, with the goatee of a diplomat, nattily and well-dressed, Vešović gave the first impression of being a man of nerves and muscle, who reached decisions too hastily and could not stand to have anyone over him, let alone anyone who trampled on his honor. He distinguished himself in his charges on Bardanjolt at Scutari, where he commanded troops of his own Vasojević clan. Some success was achieved at the price of bloody and reckless attacks, which he, as ranking commander, personally led. He went before his disintegrating unit with sword unsheathed, shouting encouragement to his men. A bullet pierced his cap, and another broke his sword. But he did not stop for a moment. For this fearlessness, which everyone recognized, the people were not grateful, because of the senseless and heavy losses, which brought ruin especially to the Vaso-jević clan. His army faltered, decimated as it was, and he was forced to use a sword to drive officers to the front lines. However, he could not restore to his army the confidence it had lost in earlier senseless defeats and massacres.

What kind of commanding officer Vešović was is shown by his rebuke of an officer, from a distinguished family at that, who was afraid to lead his unit across a defended river. The General took the officer by the nose with two fingers, drove his horse into the water, and thus they crossed to the other bank leading the army.

This, then, was the man who, when Montenegro capitulated, was chosen to await the Austrians as chief minister, to try to salvage what he could for Montenegro. The entire government, or what was left of it, was at that moment in his hands. The King and most of the cabinet had fled to Italy. They had not led the army out with them, as the Serbian government had done, though it would have been easier for them to do so. Vešović, two ministers, and the King's son Mirko were designated to negotiate with the Austrians. No one knew, because of all this, what the King and government wanted, particularly since they had disbanded the army, proclaiming that it no longer existed and that now there was only the people.

It was in that hour of general collapse, despair, and chaos that the renowned Albanian hero Iso of Boljetini fell upon Podgorica. Iso had been noted before as a rebel, a Young Turk, and a favorite of King Nikola, with whom he took refuge when the revolt failed.* He enjoyed a great reputa-

* Before the First and Second Balkan Wars, the Young Turks in the provinces of the empire rose against the Ottoman sultan but failed to depose him. Albanians declared themselves independent of Turkey in 1912 but were occupied by Serbia during the Second Balkan War in 1913 and again in 1914–16 during World War I.

tion among the Albanians, and the Montenegrins, too, treated him with an apprehensive respect. During the collapse, he abandoned his friendship for the King and for Montenegro and decided to exploit the disorder for his own gain. It must have seemed all the easier because Podgorica had no organized force to offer him any resistance. If nothing else, he might hope to plunder the city and state treasury. Terror, looting, and chaos reigned everywhere. Iso could not have known that there was still a unit that had not been disbanded—my father's. As soon as it arrived, it was commanded to maintain order.

The flying squads were generally the most experienced part of the Montenegrin army. They were, in fact, a military police force, and they received the designation "flying squads" from the big winged eagles they wore on their caps. They were charged with keeping domestic order and, like all new police forces, they were distinguished by their determination, cruelty, ruthlessness, and cunning. The opponents of the despotism of King Nikola and his camarilla had cause to remember all of them long after. But no one denied their bravery and sacrifice in war.

The battle with Iso's irregulars did not last long, despite the Albanians' wild heroism. The blow struck down both the leader and his most devoted followers. Iso's immediate entourage was wiped out to a man, and the rest scattered. Iso Boljetini himself was killed. But he had fought fiercely and long when he was left alone on the open road. Wounded, he rose to his knees and, though too weak to hold a rifle, he fired a pistol, at least to take a life in exchange for his own. Father hurried toward him, and the wild Albanian leaned his pistol on his left arm. He did not have time to fire, however. A soldier had him in his sights and—he fell. Father ran up, and Iso glanced at him with big bloody eyes, said something in his own language, which Father did not understand, and breathed his last. Father took his large Mauser, with its silver-mounted handle, and kept it as his most precious souvenir.

It was little wonder that we children mourned for Iso Boljetini. Father mourned him, too, though he was proud that his group had felled him. It was a special kind of sorrow, rather admiration for a fearless hero of wild Albania who had fought to the end on a bare field and empty road, neither begging nor forgiving, upright and without protection. There was this admiration in our sorrow, too. If one has to die, it would be good to fall like Iso Boljetini. Let it be remembered, at least by those who have seen and heard it.

Much later—we told Father and teased him about it—we read that Iso Boljetini had just died in Scutari. Father denied it. But for him it was not very important whether that had been Iso himself or one of his lieutenants; the main thing was that the Albanians who fought that battle, and especially their leader, who could have been only Iso Boljetini, were killed. Father had been told that it was Iso, and that was enough for him—the fact of his death eternally proven by gunfire.

There was still another personality out of Father's war experiences, just as wild and heroic as Iso Boljetini. But this one was from Montenegro. We knew him better from stories than if we had seen him. He was Dušan Vuković, Father's lieutenant in the regiment, and a native of the Katun district. He and Father became unusually good friends. Dušan distinguished himself by his indomitable heroism and quick wits. He was a man of great endurance, like all men who are just bone and muscle. Though we had never seen him, he remained for years in our house as a close acquaintance, known for his reasonable and thoughtful bravery, his comradely loyalty, tender manly love, and firm convictions, which he rarely changed, ever ready to lay down his head for them. Of the old stamp, he combined the heroism and ideals of his forebears with a rough cunning that the new times required.

For years, Father never ceased speaking of him, and dreamed of how they would meet again. But life would not

have it so, and they never saw one another again, not even
to refresh their war memories.

Dušan fled to the forests after Montenegro's incorpora-
tion into Yugoslavia in 1918, for he was an unyielding
adherent of the old regime of King Nikola.* My father,
however, took service in the new government, even though
he was not enthusiastic about the new state of affairs. As an
Old Montenegrin, and a Katun man at that, Dušan derived
strength for his unyielding beliefs from the centuries-old
struggle that his district had waged against the alien, a
struggle that gave birth to the freedom and statehood of
Montenegro. No matter how deep these roots were, they
were already withered; under the new conditions there was
no longer nourishment for them on the rocky soil around
Cetinje's Mount Lovćen. But in withering, they became
all the harder and tougher.

Unable to maintain himself further after three or four
years as a guerrilla, Dušan emigrated to Italy via Albania.
His King was no more among the living, while the King's
heirs had renounced their rights to rule Montenegro for a
fat sum from Belgrade. But Dušan did not renounce the
idea of a separate Montenegrin state, not even when, on
returning from Italy, he had apparently settled down in his
own district. It seemed, perhaps even to him, that the
struggle for his idea had come to an end. But it was not so.
Quite suddenly the fight flared up again, and he ran into
the thick of it, with all his odd assortment of petrified
conceptions and methods. During the Italian occupation in
the last war, though quite old, he was one of the toughest
opponents of the Partisans. Of his dream of an independent

* From late in 1915, when he fled to Italy, until his death in exile in
Paris in 1921, King Nikola hoped and planned for the restoration of the
kingdom of Montenegro, or failing that, for the creation of a Petrović
dynasty in Serbia-Montenegro (his daughter was the wife of King Peter I
of Serbia). He was deposed by a national assembly in 1918, when Montenegro was united with the new "Kingdom of the Serbs, Croats, and
Slovenes" (later called Yugoslavia).

Montenegro there remained but the empty hope that help
might come from Mussolini's black legions. When Partisan
units freed Montenegro, Dušan had nowhere to go: Italy
had already capitulated. So he tried to return to his old
trade of rebellion. He was again one of the sliest and most
persevering rebels. He had not forgotten his tricks nor lost
his ability. But the times had changed and his adversaries
were not old. His faithful supporters were no more, it
seems, and he was soon killed.

I made inquiries about him while ambushes were being
laid for him and posses sent out. No one suspected that I
asked out of an unhappy curiosity, to discover whether
anything remained of my childhood picture of him. Yes,
practically all of it had remained: his heroism and capability, simplicity and loyalty. But the picture had changed.
All these former virtues were transformed for me into their
opposites—into deceit and cruelty, treachery and corrup-
tion. Of course, my childhood vision of Dušan was, for
these postwar times, unreal. That earlier one was real and
beautiful. This new one, quite different, was real and ugly.
Yet this one could not darken the other: one was a different
man from the other, though he had survived into the present without changing. The time was different from that
earlier time.

Dušan might have known, and he did know, that the son
of his former best friend was now an enemy. But this could
not have troubled him much. Standing like a rock on a path
he had chosen in his youth, he could have regarded my
stand as the perversity of the younger generation and their
disloyalty to the ideals of their ancestors, as he understood
them. Such reflections of mine could change nothing in the
attitude of the new government toward Dušan and his
band. On the contrary, they merely incited it and me to a
merciless settling of accounts with the remnants of a time
long past. To settle accounts with the past, even with my
own, and even with the innocent enthusiasms of childhood,

seemed to become the most precious and most fiery passion of life.

These unseen but known personalities—the Czech, Boljetini, Vuković, and the rest—and various scenes from the war, however bloody and harsh, brought at least some light into our life. All the rest seemed dark and intolerable. There lay on everyone the black yoke of shame and betrayal. The undaunted and the undefeated had laid down their arms, because of the nefarious interests of the government and crown and their wooing of the enemy. All the soldiers, and particularly the officers, felt that they had been sent off the field of battle while the battle was not yet done, so that they might eke out their lives in peace beside their plows and their women. The misery, hunger, humiliation, and slaughter all now seemed senseless and in vain. The army and the land had been thrown into bloodshed and devastation just so that the petty profiteers and careerists of the court could, in the end, lead them to surrender at the price of a truce.

One could observe, in the humiliation and shame, how from day to day the Mojkovac massacre grew into something grand and illustrious, because of the blood that had washed and preserved the honor and the name of Montenegro. The blood was still smoking on the hills around Mojkovac when the legend was born of the indomitable people of these black hills, these children of the forbidding Montenegrin crags.

The proponents of good relations with Austria lost all support. From the very first day of the occupation, however, in the general dissolution, there appeared the informers and whores, drawn from the same crowd from which only yesterday the despotic regime of King Nikola had recruited its paid supporters and its last strength. On the other hand, the supporters of unification with Serbia seemed jubilant. It had been demonstrated that the despotic rule of King Nikola could only lead the country to a shameful disaster and betrayal of the hallowed ideal of a South Slavic union. Still their jubilation was not malicious; they did not rejoice at the misfortune that had befallen their land.

It was strange how, after the fall, the living began to regret that they had not lost their lives. Everyone wanted to do something, something big, but they lacked the strength, as though accursed, their minds and powers immobilized.

It might have been pleasanter in that time to be a child and not to feel the despair and the doubts that wracked the adults. But it was impossible to be a child; not even the children were spared. The events pulled children, as well, into the dark and bloody circle.

Worn out with itself, Montenegro floundered, groaned, gasped, and perished.

In the beginning the Austrian occupation was far from what might be called brutal. True, the Austrian authorities surprised even the most ardent proponents of reconciliation with Austria by their unbending attitude. Prince Mirko, the King's son, was interned, the representatives of the Montenegrin government were removed, and their officials dismissed. The regime that ensued is common to occupied territories, but it was also one in which severity was accompanied by the orderliness of an old and experienced state. But despite this, the tension in the relations of the Austrian command with the people constantly grew; everyone felt an inner necessity to wipe clean the blot and shame for which no one felt responsible.

It seems most strange that the very man designated as a liaison to bring peace between the old regime and Austria was the first to come in conflict with it. This was the officer of the capitulation, Brigadier General Radomir Vešović, who on order of the Montenegrin high command had dismissed the army, sent the soldiers back to their homes, and peaceably surrendered all power to the Austrian army. He was unable to reach any kind of agreement with Austria, which simply occupied Montenegro without regard for the Austrophile tendencies of the Montenegrin leaders. So he retired to his district, the region of the Vasojević clan, as did the other officials to theirs.

The Austrian command began to search Montenegrins for weapons. There were guerrillas in the woods, though not many and mostly youths, adherents of unification with Serbia and the other Yugoslavs, men who regarded the struggle against the dual monarchy as a revolutionary one and a sacred mission. The Austrian command had apparently decided on sharper measures, for it sent a patrol to bring in Vešović. But this temperamental and insulted man would not surrender. He killed the officer sent to arrest him and skipped off into the woods. He thereby stirred Montenegro, and particularly his Vasojević clansmen.

It was this, in fact, that brought on the really severe occupation—arrests, internment, and hangings. Only then did Montenegro bewail its lot and weep over its sons.

There was no real resistance, or else men would not have allowed themselves to be taken prisoner by the thousands. Vešović had fired the first shot, an act that might have been called heroism, but precisely because he was considered the supporter of a surrendering regime, his deed failed to bring clarity to the peoples' attitude toward the Austrian army. The confusion had not lessened; indeed, it grew.

Vešović himself issued no call to battle or to revolt; he avoided meeting the Austrian army either way. Later, he became reconciled, returned from the woods, and began to call on others to surrender. He was charged during the unification with wanting to bring the Italian army into Montenegro. But he always comported himself with courage and composure. In the course of two or three years he was transformed several times in the public mind from hero to traitor. This was like him, and like those whom he represented, and above all like the stormy and shifting times in which he lived.

It was spring and a sunny morning in 1916.

Father was busy at something in his room. My uncle's son Peter, already a lad, flew in and shouted that the Schwabs were coming. Father grabbed a revolver and threw it out the open window into a thick potato patch to conceal it. Then Father stepped out in front of the house to meet the Austrians. There were three of them. One was as orange as a fox and had a long fox's snout. He told Father that he was under orders to escort him to the command post in Kolašin, supposedly to give some information. All of us at home already suspected, knew, that father would not return. But nobody cried. Our pain was cold and full of hatred and scorn.

Father got ready quickly, as though he were in a hurry to escape the tension that reigned about him. Perhaps he was thinking of acting as Vešović had done, rather than surrender. Maybe he was afraid of bursting into tears. He kissed us all, lifting us, kissed his mother's hand and waved to his wife, and then jumped on his horse. Peter, our first cousin, went with him a part of the way.

Neither of them returned. The Austrians kept even the gelding. The house was left empty, without Father, and we felt like orphans. Father was good, and handsome. We could not mourn him enough.

My older brother, Aleksa, whom Father had captivated by his constant attentions, wept inconsolably long after. Even at night he sobbed uncontrollably in his sleep, powerless to calm down even on awakening. I would find him under the ash trees in back of the house, sitting alone on a rock, wiping his cheeks with thin bony fingers, while the tears ran and ran inexhaustibly from his large, bleary, nearsighted eyes. His big, transparent, and cold ears were so sad then. I, too, wept, not so much over Father as over my brother. Father soon became a shadow—a substantial and real shadow because he was talked about at home—but one without warmth, whereas my brother's tears were tangible and inconsolable, somehow a part of himself. Our younger brother would join our weeping, even though he did not know what it was all about. He wept noisily, bawling, but with just as much sorrow, even more sorrowfully, for he wept only out of grief for his brothers.

Now that Father had been led away to a distant foreign land, he became even more real and closer through his rare letters, written in a legible hand, and the little packages Mother would send him after saving up rendered lard for months. Now he was someone whose coming we awaited, and thus he became dearer and more precious. His picture was placed on the wall, only his, and it reflected both our sorrow and our hope. He was as though real in that picture, sometimes stern, at other times engaged in amusing conversation, all cleverness, or tawny and hunched up like a jackal in a forest glade.

Thousands of Montenegrin fathers and youths were led away in the same manner. But this affected the resistance of the people less than the cowardice that the court and government had shown previously. On the contrary, with this the peoples' resistance to the Austrians really began, this and the Austrians' acts of violence and brutality. First they hanged three Montenegrins on a bare hill overlooking Kolašin. One of them had been educated, and was charged with inciting resistance. The second was Radomir Vešović's brother. The third was an aroused and unsubmissive peasant. The black shadows of the corpses on the high gallows lay across every soul throughout Montenegro. Years have gone by and still these gallows reach heavenward, in memory, while at the foot of the hill a beaten and terrified people keeps silent.

After that nobody could trust the Austrians. Everyone could expect to be interned or mistreated, if not strung up first. Soon all came to know the names of those Austrians who committed the worst acts of brutality. But the names of the guerrillas who comforted themselves and the people were also known. In our district two such Austrians were known by their ill repute—Krapež and Bilinegije; and two rebels by their good repute—Nedeljko Bošnjak and Todor Dulović.

Todor Dulović was a distant relative of my father and, because our house was in an isolated spot, he came by fre-

quently. We children were constantly reminded, though
we knew our duty without this, that the arrival of guerrillas
must be kept secret from everyone. We were especially
cautioned to guard against being tempted and enticed by
presents, sugar, or figs. The adults were afraid of our child-
like naïveté. Actually, we children were more cautious and
discreet than our elders, for we felt a special charm and
pride in keeping a secret.

Our house was never a part of the underground, not even
during Austrian times, much less later. We delivered no
messages to the guerrillas, nor did we hide anything of
theirs. But they knew that no one in our house would
betray them, and that they could always obtain food there.
It was the same with many Montenegrin homes, even with
the majority. There was a little-traveled road by our house
to Mount Bjelašica and the vast Biograd Forest, King
Nikola's stronghold, and there rebels often found a sure
haven. Our house was a convenient place to rest and refresh
themselves, being concealed and safe.

Todor was distinguished above all the other guerrillas
not so much for his bravery, for he was not overburdened
with it, as for his good sense and, especially, his manliness.
He was capable of going hungry for days without taking a
crust of bread from anyone unless it were given to him
voluntarily. This distinguished him all the more because
there were many who plundered our people, too, and for
whom the life of a guerrilla was a convenient way of making an easy and good living.

Todor became famous quickly during the occupation
itself. In the war many soldiers had put a better foot
forward than he. But under the Austrians he was among
the first to take to the woods, bringing wrath on his house,
but giving no thought to the persecution his family would
suffer. He soon became noted for the boldness of his attacks, for which bravery was not enough, as in war, but
which demanded great skill. In war the command helps the

soldier. Now Todor had no one to help him, nothing but his head. He had quite a head, one more audacious and wiser than others, and, moreover, more honest than the rest, and therefore he had been accepted as a leader and recognized as one of the most distinguished rebels in all Montenegro.

Todor was not a handsome man. He was rather short, thickset, and his face was very swarthy. He had a thin little mustache, a square low forehead, snub nose, and a thick hanging and split lower lip. His nose was congested, and he snorted through it so loudly that one could hear him several paces away. His snorting was particularly loud when he slept. There was something grotesque, and at the same time something good-natured, about him. He never took off his weapons or put them beyond reach. He did not like a big band, only three or four, at most five or six, but men tried and true. He never laughed much, nor was he much for talking. He was terrible only at first glance. He liked to play with children, and did not even keep them from going through his ammunition and weapons. Toward the old people he was attentive and obliging. He was the soul of simple courage and goodness.

As soon as a soft and persistent knocking was heard at the door in the night, everyone in the house knew that it could be no one but Todor and his band. Black and sweaty, he would sit beside the newly stirred fire and wait with his band for something to eat.

The black night there received him, the black one, as it had brought him.

Nedeljko Bošnjak sprang into my memory from a wet, warm summer twilight just before the end of the war, though I had known all about him before. A shy shower had just flitted by, leaving everywhere in its wake the dewy grass and the wet thick leaves of the forest. A whistle pierced the dusk. There could be no doubt but that it was a guerrilla. Something black and bristled slid out of the

dark thicket across the thick green meadow. The monster
loomed larger and slowly approached. It was Nedeljko in
a mangy cloak of goat's wool with a hood. Behind him
trailed the night out of the mountains. He smiled like an
old acquaintance, with big white teeth in a thick brown
beard. He was of middling height, more stocky than
slender. He was still a young man, very good-natured and
shy. When he discovered that only the folks were at home,
he came, barefoot, his boots under his belt, walking up the
path as though he knew it well, and helping to chase in
one of our calves as though it were his own.

Nedeljko did not visit us often, but he always came suddenly, in the dusk. He would also help with some of the
household chores, even splitting the wood. He did not seem
fitted for the forest and guerrilla life. The village liked him
as a good though rarely seen friend. The war had brought
him to our region all the way from Glasinac, in Bosnia, and
this is why they called him Bošnjak. He had deserted from
the Austrian army in the very beginning of the war, and
had come to Montenegro as a volunteer, staying after its
fall. He hid under the open sky until the notorious Krapež
met him in Mojkovac, slapped him around, and arrested
him. Somehow Nedeljko slipped away. Now the Austrian
authorities were looking for him, as their subject, and as a
rebel in addition.

Everybody knew Nedeljko was stalking Krapež, to kill
him as revenge for the public slapping. At least that is what
our Montenegrins thought as they urged the Bosnian to
take revenge. But Nedeljko needed no urging.

Among Montenegrins it was a rare thing to find industry
and a warlike spirit in the same person, but the Bosnian
helped the women in their work as easily as he wielded his
weapons. He liked to talk, but just about his village, the
cattle, and his family. His speech drawled and was soft,
and, most unusual of all, he would throw in swearwords,
not as insults, but simply as a manner of speaking. With us,

however, one either swore or one didn't. Though a chatter-box, Nedeljko was no braggart. He never vaunted his heroism. This did not in the least detract from the bravery of this man, not even in the eyes of the Montenegrins, for whom something becomes not only great but remarkable only if it is sung about and praised. He achieved the reputation of a hero after some time, distinguishing himself by his cold calm in tight spots, as though he were about his ordinary business.

Once two Russians showed up at our home, prisoners who had escaped from the Austrians. It was rumored that there were Russian prisoners everywhere. No one wondered much at this; after all, there were enough Russians to populate the whole world. These two apparently were jacks-of-all-trades, though in our villages there were no other than jobs for masons and carpenters. Finding nothing to do, they insisted on repairing shoes. There were no shoes at our home or, for that matter, in the village either. But they took Father's boots, which were already stiff and misshapen, and put them in order, even though there was no one to wear them. It was a matter of doing something, at least, for a handful of food and a night's lodging. Most remarkable of all, the Russians, too, wore peasant sandals. Maybe their lightness made it easier to flee. Everyone liked the Russians with a rather sad love that they themselves evoked with their constant tearful blessings. These Russians were hungrier and more wretched even than the Serbians. Their homeland was farther away. One of them, small, dark, bloated, his face and hands all gnarled, tenderly caressed the children, probably out of yearning to see his own. He gripped the boots between his knees and banged on them so hard while repairing them that it seemed he would nail them to his knees. They stayed two or three days, and then got lost in the crowd. Where did they go? Where were they from? Were their homes in an empire in which, they say, the sun never sets?

Other guerrillas came in addition to Todor and Nedeljko.

Once in the middle of the night in the summer of 1918 came a whole gang of unknown men. They had stolen a herd of cattle from the Moslems across the Tara, and were driving it to their own villages, toward Rovči and the Morača. They came to stay with us only for the day. While it was still dark they drove the cattle into the woods a distance from the house. In the morning they slaughtered a calf, whose blood spurted over the dewy grass and damp rocks. We children had a fine time roasting pieces of entrails over the coals beside the dozing and sleepy men.

We did not know whether men such as these were guerrillas or just plain marauders. They would stalk cattle by night, along the side roads, leaving perhaps a head or two with people they trusted, to make the herd as small as possible in case they should encounter the Austrians. Later, they would gather together the cattle they had left behind and sell them, splitting the profits among themselves. Thus there came into being a clandestine black market in cattle stolen entirely from Moslems. To be sure, the real guerrillas never engaged in such dealings, only brigands did, but there were just as many of them as of the others.

Once, while they were bivouacking in the woods, I found an opportunity to snatch a grenade from one of them. As I unscrewed it, it slipped out of my hands and fell so that the fuse struck a rock. It ignited easily and began to sizzle. The guerrilla, a ruddy and stocky lad, grabbed the grenade quickly and hurled it down a ravine, then fell flat on the ground clutching me tightly. All the others fell, too, as though the blast of the grenade had flattened them to the ground. Though no one was hurt, they continued to lie there for a long time, as a shower of dust rained down upon them and tattered leaves floated in the air.

Among those who came with cattle was an elderly man with a large mustache, a pistol and binoculars, with bandoleers and feather frills—all decked out. They said he was the chief. Not understanding the full gravity of my words, I remarked, "Huh, steal a cow and be a chief!" The story was long told and retold laughingly, how somebody had actually dared tell the chief what he really was. I understood the laughter, though I could not comprehend why both at home and in the village my words were regarded as a sign of great bravery. He was really a chief of cows, of stolen cattle. Because he was this, it seems one did not dare tell him so. It is considered all right to tell others pleasant things, but not to tell them the truth. Bravery means to tell the truth, in most cases, it seemed to me.

Always playing with weapons, frequently in the company of armed men, the older children would steal weapons, which had been hidden away, and shoot at Austrian patrols from thickets and hills. Not only the villagers but the guerrillas as well got after such children, for they only brought trouble on their necks. The children had become so wild that the villagers would not risk chasing after them; they would wait until they came home and then beat them.

The brigands plundered Montenegrins as well as Moslems. Once, in summer, Mother was milking the cow. The spurts of milk gleamed white in the dusk. Three armed men rushed suddenly from three sides of the meadow and surrounded the barn. Mother stood up and one of them spoke softly to her. Why softly? She began to protest loudly; these would-be heroes could do as they pleased— to women and children, that is. In addition to our cow and calf, we had a young bull, Spot, all white except for a patch around the middle, a good-natured and quiet beast. They tied Spot without haste, as though he was theirs, and began to wend their way into the darkness. Then Grandmother came to curse them as only she was able to do, with curses transmitted to her through ages past. Her torn blouse came unbuttoned and her withered breasts fell out, and her white hair tumbled down. The men hurried away, almost fleeing from her, and soon night fell over all, black rain began to pour, and lightning illumined the sky. Terrified by the darkness, the thunder, and the men, we huddled trembling in bed around Mother, while she lamented the entire night.

Still earlier, brigands stole nearly all the potatoes from the field and even the unripe fruit from the plum trees. They came by stealth, while the other men had come in plain sight to take away our bull, as though they had a right to him, as though we were enemies, Turks, Schwabs, or spies.

Nevertheless, with us children our greatest fear was not of men, of brigands, or of Austrians. That fear became mixed with another—the fear of nocturnal apparitions, of evil spirits who were everywhere and could appear at any time.

In the hills around the house foxes barked and dogs bayed in the dark. Their sound was thin and enticing. The forest rustled all through the night. The trees conversed with the hilltops, the branches creaked, an owl signaled from the crags, and squirrels joined in from a distance. Perhaps these were all devils who lived in caverns and led Christian souls astray. A monster with a thousand damp hairy hands and huge claws was abroad and lay in wait in the surrounding woods. It held the house in its clutches waiting for someone to venture out. When the moon emerged, white apparitions, long and vast, waved among the clump of trees beside the meadow like vampires in white robes. And that ceaseless groaning of the Tara and the churning of the whirlpools were also devils who arose in the wee hours, chattered and howled, stuck out their big purple tongues and flapped them against the wet rocks. The sprites played with the leaves in spring, and those who belonged to our clan, our grazing land, waters, and hills, fought with the sprites of other clans. One had to be careful whenever a whirlwind arose that no leaf was punctured, for this meant a sprite, maybe one's own, had lost an eye.

It was necessary to flee from a whirlwind, to avoid having its shaft pierce your heart.

Only in the stars, during clear nights, was there joy and freedom. The heart took courage at the sight of them.

Grandmother believed most fully in all these apparitions, in fairies and dragons, vampires and sprites, witches and demons, spirits and werewolves, as though she had spent all eternity with them. She knew them, felt them, saw them. Mother would say, "There is nothing to it; it is all a story." It was obvious that she said this in order not to frighten us, yet she, too, believed.

And how could one not believe? Grandmother knew people to whom things had happened, and she saw much with her own eyes, like the black dog who ran around her legs and mockingly tried to lure her away from the road until she crossed herself three times. She knew, by name, any number of witches and men who had become vampires. Whenever one of our animals was lost, she would step outside the house in the middle of the night and cry aloud, calling upon the souls of the drowned and on good spirits to watch over our cattle. When she called out into the night like that, beside the mountains, everything seemed to come alive—the woods and grass and stars—all straining to hear her and wondering what to do. Most of all, she called on a drowned person known as Golub Drobnjak. When had he drowned? Perhaps when her grandmother was still a girl. And now his soul was wandering over the earth, unshriven but just, seeking peace.

As one walks or sits, sleeps or eats, the world of dreams mingles all together—men, Schwabs, and guerrillas with phantoms, things with designs. As one grows, as one attains consciousness, this world of dreams does not fade but spreads and deepens with every thought and feeling. Perhaps life would be more closely bound to that world of dreams if constant strife did not remind us and call us back into the coarse world of things and men.

Everything is at war with everything else: men against

men, men against beasts, beasts against beasts. And children against children, always. And parents with children. The guerrillas fight the Austrians, and the latter persecute and oppress the people. The spirits strive with humans, and humans with the spirits. The strife is ceaseless, between heaven and earth. And Mother beats us. If she cannot catch us during the day, she beats us when we are asleep. The switch cuts into the flesh, and one sleeps on. And when we awaken, she demands our promise that we will never again do what we did. Or else the beating is continued.

It would be easy to promise that we would not do what we did if only we could feel truly guilty for what we did. But since we do not feel guilty it is better to lose some sleep and endure the beating to the end.

Certainly strife is one side of life. But there comes a time when only strife is the order of the day, as though there were nothing else in life.

My mother sent me to school while I was still so young because, so she said, I was such a bad child. She could not stand me. I had just then reached the age of six. Other children began school at a later age, some even past ten.

The school was a full hour's distance away, and consisted of a room in a peasant's house, because the real schoolhouse had burned down during the war.

The year was 1917.

In this very year there was a drought that can never be forgotten. A frost and then a drought destroyed everything. Even had there not been a war, hunger would have invaded us. People ate wild herbs and sawdust made from beechwood. In the spring Mother gathered nettles and placed before us unsalted green globs of it. We had a cow that we loved as a member of the family. Whenever some sour milk was added to the nettles, and especially if a bit of flour was mixed in, this was a much-awaited feast. We were hungry often. Famished, my brother and I bewailed our fate and once, while weeping, fell asleep, our arms about each other, on the path that led across the meadow from the house to the spring. It was then for the first time that we spoke of death. We would die of hunger and would be buried. We would never see one another again.

Deaths from starvation were not rare. The child of our neighbor, Toma, had already died of hunger. He was a handsome little boy with big sad eyes, which seemed even bigger and sadder to us after his death. They were all that remained after him. Though always starved, he never seemed voracious; he would only stare vacantly while someone else was eating. My brother said that we could have helped and remembered occasions when we had not. The entire village took his death as its own grave fault; the child might have been saved somehow. Nobody talked about it, but everyone felt as though he had killed the boy. All expressed horror at greed and selfishness, yet all were greedy and selfish. Fear of starvation is frequently stronger with those who have something than those who do not. Those who have talk about how they have nothing. The others keep proudly silent and endure, as though they have some other great wealth, but wealth that cannot be eaten.

The cattle starved, too. The pastures turned into wasteland, and the springs ran dry. To find grazing land, and also to protect the cow from brigands, we took her in the summer to Father's standard-bearer, Radivoje Adžić. Radivoje was a rather stocky man with a large head. It was his manner of speaking that distinguished him from others—brusque, sporadic, boastful, and confident. He had a large family settled right on the cleared bank of the Tara. They were poor people who had no house; they lived, both summer and winter, in a shack. Yet Radivoje held himself proudly, as though he were in the thick of a battle, holding aloft the flag which not even death would force him to let out of his hands. In his poverty, the flag and pride were all that remained to him of his forebears.

His family was poorer and hungrier than our own. Large man that he was, he could never get enough. In the evening when it was time to sup warm milk, he and a herd of children surrounded the bowl, he with a huge spoon, a wood spoon, for he never liked to eat with a steel one because it would clatter so against his teeth. In the scramble for the milk he would say over and over, "Easy, children, take it easy. You'll eat it all up and won't leave anything for the devil himself, devil take you." And he would slurp faster and faster. Then, with all that liquid in his stomach, he would go to guard the corn from the badgers. He would shout all night, and roast ears of corn on the sly. In the morning he would say that the badgers ate it all; bitter starvation pressed them as well. What could one do? But his resting place, under the lean-to, was surrounded with corncobs, gnawed and scorched.

In all of this there was something both funny and sad. The standard-bearer was a hero, perhaps less of one than he boasted, but a hero, nonetheless. He was a man of the old stamp and proud, even though hunger drove him to push with the children for milk and nettles, to pilfer from his own family, and to blame the badgers for his own theft.

Violence of all kinds increased daily.

An Austrian sergeant attacked a girl in the neighboring village, in Bjelojevići. Her brother, Manojlo Mišnić, flared up and killed him. The Austrians burned down the village, and the whole mountain was veiled in green and acrid smoke. The winds bore the stench of the smoldering ruins. Manojlo hid on Biograd Mountain in the secluded hut of a widow named Marija. She betrayed him, so the story went. One dawn the hut was surrounded. The Austrians called to him to surrender, but the outlaw would not dream of giving himself up. He threw a grenade at them and then rushed out after it. He might have escaped; he was on the edge of the forest when a bullet struck him in the heel. He was overtaken and felled, cut to pieces with Austrian bayonets. The peasants dragged him to the meadow overlooking Biograd Lake and buried him beside a clear cold spring where travelers were wont to rest. They chose a beautiful spot, where his grave and death would be remembered. But all was forgotten just the same, covered by the fern and submerged beneath the weight of more memorable events.

The guerrillas did not wait long for revenge. They tore out Marija's tongue and killed her. Though Marija had

been a hungry woman with hungry children, she was
guilty of betrayal, and even those who pitied her did not
consider that an injustice had been done in killing her. But
few approved of the torture, which was like the torture
by the Turks of outlaws described in their popular epic
poems.

Whence came such sudden cruelty?

In the house next to the school, guerrillas planned vengeance on a peasant who was suspected of being a spy for the
Austrians. They called him one night, but it was his sister
who dressed and went out, and in the confusion they shot
at her. A bullet crashed through her teeth and came out at
the base of her skull. Had her brother been killed, probably
no one would have mourned him. The girl had not been
pretty, but she had been strong and good-natured, and for
two days there were no lessons in the school in mourning
for her youth, mourning that was all the greater because
she was a maiden. She was innocent—and this made it all
the sadder.

Another woman from a distant village, a noted beauty,
went around with the Austrians. She was stripped naked
and crucified at a crossroads. This especially evoked real
horror. How could anything like this happen in Montenegro? People said that a harlot should be stoned or hanged,
but not this.

There was in the neighboring village another pretty
woman, with catlike green eyes and black hair, a young
widow, all soft, cuddly, and white. She stopped by often,
either to visit us or our godmothers. Childless, she loved
children, delighting in caressing them and never forgetting
to bring presents. My mother avoided her, and it was obvious that she did not like her because of her connections
with the Austrian gendarmery commander. He was a man
who was, by the way, easy on the people, and a Croat
besides, who not only avoided pursuing the guerrillas but
gave them warning of raids.

The children liked the lady with the green eyes; it was as though they did not wish to know of her sin, or as though they hardly suspected. She was so good and kind that she was not punished severely. They shaved off her beautiful thick black tresses—took the crown from her head—and shame would not permit her to go among the people any more. She was guilty, they said. She could not hold out against want, youth, and beauty. She never came again to embrace us. She simply vanished, with all her beauty and shame. Her lover did not take her with him. In the memory there remains only the warmth of her embrace and the hardness of popular justice.

Todor set fire to a lumber mill that had been working for the Austrian army. Unfortunately, it had provided a living for not a few poor people. He killed both the guard and the Moslems who worked there. The deed was a heroic one, committed under the noses of the Austrian army, but bloody and cruel beyond measure.

It seemed that everyone tried to find some moderation in revenge, but could not find it, and so were goaded all the more.

So it went. Spies were punished, Austrian soldiers were killed, property was razed, Moslem cattle and house furnishings were looted. And from the other side there were reprisals. War and hunger continued, coupled together, conceiving and spawning fresh violence and crime.

We Montenegrins did not hold a grudge against the enemy alone, but against one another as well. Indeed, our enemy—the Austrians and their minions—were called to intervene and to help in these quarrels. Two notable clans entered into a blood feud. No one really knew what it was all about. While one side did their shooting as guerrillas, the other side joined the Austrians. The Austrian shadow hovered over all these crimes. But the root was in ourselves, in Montenegro.

Nobody could really escape running afoul of the Aus-

trian authorities. In our villages that meant the notorious
Krapež. I was going down the hill one day to the village;
the dirt road had become furrowed and slippery after
a summer shower. Krapež was walking below. Everyone knew him already, and pointed him out in whispers
and signs. With him were two guards. He was a ruddy
man, as though in flames even when he was not angry.
In school we had been ordered to greet Austrian soldiers
by raising our caps. On seeing Krapež I began to hesitate
between taking off my cap and going by without greeting
him. I had time to think it over, but reached no definite
decision. I got to one side of the road, but I did not remove
my cap. Krapež stopped, stretched out a meaty and mallet-
like hand, and gave my ear a powerful twist. He said
something to the two with him, perhaps something like:
What kind of people are these, when their children are so
bad! He ordered me to greet him properly the next time,
then released me and slowly went on. I hurried off and
then furtively turned to look. Krapež was rounding the
hill. The stories about him seemed to me to be more terrible
than he really was.

Hunger and the paw of foreign rule were felt in the
school also. In order to get books, we had to gather acorns,
for which the Austrians had an appetite. But not even
these books were ours; they were theirs, the Austrians'.
My primer had a funny Croatian title. It was written in
the Latin alphabet. We secretly studied the Cyrillic alphabet from Montenegrin primers belonging to the older
pupils.* Its letters were not as pretty, but they were big
and the words were understandable. True, even in them
there were occasionally Turkish words for Slavic ones;
this was only to accustom us to a certain letter appearing

* The Croats and Slovenes, being for the most part Roman Catholic,
use the Latin alphabet. The Serbs, including Montenegrins, use the
Orthodox or Cyrillic alphabet, as do the Russians. Serbo-Croat is basi-
cally a single language despite this use of two alphabets.

largely in such words. At least the Montenegrin primer did not have those silly Croatian words. The teacher did not keep us from learning Cyrillic on the sly, and, seeing that we were starving, he did not beat us much.

I soon learned that it was not a good idea to be the best pupil. I would remember what the pupils in the upper grades were learning, and would volunteer when one of them did not know something. The older pupils then became offended, and would have beaten me had I been bigger. Children are rather more considerate among themselves than are adults. They will take care of the weak. My brother did not like my superior knowledge and teased me for it. Though I would have retired, the teacher would not permit me to. He always put pressure on me and upheld my reputation. I was aware of being the best pupil in the school. I was not proud of this, nor did I boast, but merely took it as something natural, something gained through no special merit of my own. The teacher was a young man, and not very strict. He had not been able to finish normal school and was better at understanding our misery and poverty than he was at teaching us anything. Was it only my love for him that made me work so hard on his assignments, even late into the night?

Not even the school, with its books, lessons, kneeling, beatings, play, and teacher, was able to remain a warm, self-contained world. There was the Latin alphabet, the picture of the Austrian Emperor on the wall, and from time to time the Austrians would visit the school. The teacher was obviously afraid of them and their spies, even though we children were loyal to him with a unanimity he could not imagine.

The ubiquitous Krapež once came to school. He did not interfere. He just sat and listened. As he was leaving, he frowned and whispered something to the teacher. He seemed to be looking at me. Had he recognized me? Was it impossible, even in school, to escape the war and Krapež?

In a hollow beside the village lived a very poor and
decent family, an old woman and her grown sons. The
youngest of them, Blažo, was still beardless, and blind in
his right eye. He was a happy-go-lucky sort, and his blind-
ness merely gave him a merrier appearance. He was called
One-Eye, and not by his name, just as in the folk tale.
Since his house was at the edge of the forest, it provided
a frequent refuge for Nedeljko Bošnjak. Blažo was his good
companion in everything, though younger than he. Either
it was Blažo who found out, for he went about freely, or
else some informer brought Nedeljko the news. At any
rate, the two learned that Krapež was to pass by on the
road across Mali Prepran, through ravines and evergreens.

Later, the people were to ascribe what happened to
Bošnjak. But Blažo had a different story.

Nedeljko talked him into coming along to witness the
great deed and to get a taste of Austrian blood. When they
found a spot, however, Bošnjak began to look for another.
He complained that he could not see well enough from
there, that the bend in the road was too short and that
he would not have time enough to take good aim, and
moreover . . . He even began to complain about his gun,
though he had a Mauser, while Blažo had a rusty old
Wendel with which even his grandfather had waged war.
Blažo could see that Bošnjak would like to make his kill,
that he yearned for it, but that his heart was frozen. Blažo,
a Montenegrin stripling, could not imagine letting Krapež
go by. It was now or never. A man is like a rabbit: if you
let him escape, you'll never get him again. The two finally
separated and agreed that Blažo would not shoot until he
heard Nedeljko, whom Krapež had to encounter first.
Bošnjak wanted to be the first to feel the joy. But
Bošnjak let Krapež go by. He was already some distance
away when One-Eye let him have it with his ancient
blunderbuss. Krapež fell and the guards fled. Bošnjak and
Blažo ran out on the road. Krapež was still alive and began

to beg for his life. Bošnjak, it seems, was willing. He was
that softhearted. It was shame not hatred that drove him to
seek revenge. Blažo thought differently; they had taken
on a job, and they ought to finish it. He placed a foot on
Krapež's throat and bashed his head in with the butt of
his rifle. Bullets were scarce and couldn't be wasted on the
likes of such fellows.

It was hard to find out the truth about how Krapež had
met his end. Perhaps Blažo did not really brag about it. Maybe Nedeljko had changed his mind on seeing five or six
guards around Krapež. Maybe his heart trembled—it can
happen to the best of heroes. Nedeljko was not a braggart
anyway, while Blažo had to keep quiet as long as the Austrians were in Montenegro. As soon as the occupation was
ended, Bošnjak went back to his own province. And so the
glory fell to him, while Blažo got Krapež's silver watch.
One can never learn the whole truth about anything.

Another who was killed was Ferjančić, also an Austrian
officer. A song about his death was already being heard.
He had killed a girl who was in league with her rebel brothers, and they ambushed him the very same day and re-
venged their sister.

In this land a man may not get the punishment he
deserves, yet he can never quite escape. So it was with
Bilengije, who was even more notorious than Krapež. He,
also, was downed by a rebel bullet, beside Umukli Vir, a
mysterious quiet spot where Saint Sava * once silenced the
Tara with his staff so that it would not interfere with a
pleasant conversation he was having. There, on the spot
where Bilengije fell, a yellowish faded grass grows, faded
with human blood. He, too, was a man, though an evil
man and of a foreign people. No monument was raised to
him. But the earth received his human blood. To the earth
he was like anyone else.

* Saint Sava (died 1236 or 1237) was the organizer of the Serbian Orthodox Church and the first Bishop of Serbia.

In all parts of Montenegro there were killings, some ours and some theirs. And on her borders there was more and more looting and pillaging.

This land was never one to reward virtue, but it has always been strong on taking revenge and punishing evil. Revenge is its greatest delight and glory. Is it possible that the human heart can find peace and pleasure only in returning evil for evil?

Krapež met his end in the autumn of 1918. No one even suspected at the time that it was a portent of the fall of a centuries-old and mighty empire.

The villagers did not have time to take secret joy at the enemy's death. It was expected that the Austrians would, as elsewhere, take heavy reprisals, perhaps even by burning down the homes and arresting the whole village. With each moment the horror mounted and dire prophecies grew darker.

We rushed helter-skelter to take our things out of the house and to hide them under fallen trees and rocks. We drove the cattle deep into the woods. We hid even the chickens. The house suddenly gaped empty and bare. Nothing living was heard. Only the cat mewed sadly, as though she felt the emptiness and the terror. So the days went by. How many days? We slept on blankets spread on the bare floor, all together. The night brought relief; another day had passed. Yet, the terror did not leave us. What would tomorrow bring?

One night the firing began. Rifles burst on all sides, the bombs roared, and a machine-gun barked from atop the hill. What was happening? Who was fighting with whom? No fires were to be seen. Were they shooting down everyone to a man? Maybe they were firing only to dispel their own fear? We awaited the dawn, wide awake and trembling. Then came—silence. The Austrian army was retreat-

ing. And everyone turned on it. The night again brought
the din of battle.

Both the villagers and the guerrillas attacked them from
the roadsides, trying to snatch a horse or a man from a
cart in a straggling column. They were ambushed in the
hills, in wooded ravines, and in houses. The army of a
stricken empire was beating a path through hatred and
death. They were killed on the roads, in their bivouacs at
night, at springs where the weary drank water. The strag-
glers were killed even by women and children. They had
their clothes and weapons taken from them and were sent
naked and unshod into the forest wilds, in the middle of
autumn, to be killed by someone else when there was no
longer anything to be taken. All kinds of supplies were
abandoned, as well as the unburied nameless dead, who
succumbed to fatigue or were felled by some hand. Austrian uniforms and supplies began to make their appearance
in the villages.

Two women enticed a pair of Austrian soldiers, who
were in the village, to the brook, disarmed them, told
them to sit—and then killed them with axes. Everybody
praised the women for their heroism, passing over in silence the horror of their deed. But why did the soldiers
submit so easily? Are there men who do not defend themselves against death? The main body of the army retreated,
apparently, in order and in formation, so that scattered
and unco-ordinated peasant bands did not dare attack,
except to take pot shots from the distance and harass the
flanks. But these men, left behind their units, in a land
whose language and customs they did not understand,
stranded and deserted, while their empire was crumbling
to bits, had lost every hope and dumbly resigned themselves
to fate.

Once again an army lost a war. Only this time it was
not ours, and there was no one around to mourn. Its misfortunes were far from ended, as is the case with every

defeated army, especially in a strange land. Now everyone was a hero—against a weary, tortured, and confused army. Everyone said, "Good for them! What were they doing in our country anyway?" And they were right. Yet these unknown soldiers were no better and no worse than our peasants. They were hardly to blame that some inscrutable force cast them on our land, or any other. Now our peasants were hunting down and killing Austrians; they, too, were driven by a relentless force. That was war; it was the least guilty who paid. One side lunged at the other, the very people who had least reason to do so.

As in every criminal deed and dishonor, there sounded out deep from the masses a humane voice, alone among thousands, but noble and unforgettable. There was a woman, a Montenegrin, who had no more pity for the Austrian army than the rest, but who sorrowed at the human suffering of soldiers in a strange land. She drove her husband, who had taken some soldier's boots away from him, to find the poor man and to restore them to his bare and bleeding feet. She said she did not want a martyred soldier's curse to overtake her children. Spare and bony, all bent and sucked dry, she stood before her country and her people, great and pure. Human conscience and compassion are never completely stilled anywhere, not even in Montenegro in moments of drunkenness from holy hatred and righteous revenge.

The Austrians withdrew quickly. Their misfortunes and corpses along the roadside were forgotten just as quickly. But the evil days did not go with them. It seems that they were yet to come. The guerrillas took over the government in the cities. But what kind of government was it? Peasants from the woods, without officers or discipline, many of whom were marauders and bandits. In fact, there was no government at all. In Montenegro, at least in our region, little harm was done. Men somehow became closer to one another. It was different, however, in regions with a Moslem population where power fell into the hands of the Montenegrins.

There it was as though some fury, a great fire, suddenly seized an entire region. All rose up—young and old, women and even children—to pillage the Moslems in the Sandžak. Even men who were not ever easily misled, who had lived in righteousness and meekness all their days, now lost their heads. For many it was a desire not to gain, but simply not to lag behind the rest, not to let someone else get something they could snatch just as well for themselves. All knew it was a sin to loot. Nearly all went anyway, as though afraid to miss something great and fateful. It was a kind of mass migration or a religious frenzy. Even those who opposed looting and who tried to persuade others not to go finally went themselves, for nothing could be done anyway to call a halt or to mend matters. Men came from other regions, from all the corners of Montenegro, led, it seemed, by some overpowering instinct or enticing scent.

There was much truth to the claim that the Moslems had helped the Austrian forces of occupation. There was even more truth in the contention that both they and the Montenegrins had been accustomed through centuries past to plunder one another. Yet this looting would have happened even without any of that. The Montenegrins plundered people of the same blood and tongue, but whom they considered Turks because they were of an alien—Turkish—faith and a pillar of Ottoman power through long centuries of Serbian enslavement. There was something else in all this, too, something even deeper and more lasting, a kind of perverted vow, some deep inner pleasure at attacking an alien faith with which a struggle to the death had constituted a historic way of life and thought. Now it was the turn of the Montenegrins to settle ancient scores. Who knows if they would ever have another chance? And then, too, there was some little profit in it. One does not go without the other.

Not even my mother held out, though she always used to say that stolen goods were cursed goods and that no one ever found his luck in looting. True, she never engaged in real looting. Her target was wheat. The guerrillas were distributing things from an Austrian warehouse in Bijelo Polje, and our house, which had always been hospitable to the guerrillas, had been apportioned a whole load, complete with Moslem driver. But the driver was snatched from Mother's hands on the road and killed.

Two uncles, from two different directions, suddenly made their appearance. Mirko arrived on horseback all the way from Nikšić, and on the very same day, without paying any notice either to us or to his land, joined the looting in the Sandžak. Lazar arrived some two or three days later, and he joined his brother. But they came too late for the real looting. Everything had already been cleaned out. Mirko forced a Moslem child to squeeze through the bent bars of the window of a mosque and to hand him the carpets. This was charged to Uncle as a great sin and shame, like desecrating a church, and we never talked about it in our house. His mother told him that he would bring great misfortune upon us and not to bring any of his loot into the house. But my godless uncle laughed! And so the rugs came into the house. Lazar was not able to liberate anything, and so he began to steal chickens. He brought, on the back of his horse, two crates of chickens. Above the cackling of the chickens rose howls of laughter. In addition to all the jeering, some guerrillas almost disarmed him. Unheroic as he was, Lazar bristled and shouted that they had better not get in his way, for he was Nikola Djilas's brother. So he saved himself, not by his own courage, but through the renown of his brother's name, and, armed to the teeth, he rode into the village with his load of chickens.

Many Moslems were killed. However, there were apparently no general massacres anywhere. The mountain settlements of Moslems offered armed resistance, while in the towns it was difficult to kill openly. Though men saved themselves through bribery, nobody could save himself from being looted, unless he hid something away.

The pillaging was accompanied by a terrible epidemic of Spanish fever, as we called it. The Spanish fever spread with the looting and incited, in turn, to more looting. Blankets and sheets were taken from sick and infested Moslem families, and dishes, too, and all of these articles made the rounds as gifts, loans, and restolen goods. In our village there were hardly enough people left alive to bury the dead. They had either died, or were sick, or were away plundering. The disease cut down mostly the young and strong. It was a new war for Montenegro, for those who were only now having their turn, but this time with an even more dread enemy. Whole families were wiped out. The keening never ceased.

Danger lurked everywhere and was oppressive. In the footsteps of the Austrians came the Serbian army,* few in numbers and weary. The villages filled with rebels, while a national guard was formed around the small towns, largely of youths who supported unification with Serbia. In Kolašin there was already a new government of Montenegrins, supporters of unification. The first clashes took place over disarming the peasants.

At home the sorrow was greater than the fear; we would die before Father found us and took joy in us again. Perhaps death would waylay him on some distant road, on his way home. As in other families, we became accustomed to death and had no fear of it. But sorrow, the sorrow before death was great. How many children and fathers never see each other again?

* A part of the Serbian army had escaped from the enemy's encirclement in 1915 and reformed on the island of Corfu. In 1918 the Serbs and their allies attacked the Central Powers' forces in Montenegro as part of the Allied drive that embraced the Macedonian and other Balkan peninsula campaigns.

Then Father returned, even more spare and gaunt and
graying than when he was taken away. He was still nimble,
like an old wolf, from afar, hungry and tired. He brought
with him the snow, thick and cold. The unexpected joy
petrified us. My brother could not move his head from
Father's breast, where he had buried it. And Mother, to
hide her tears—for it was not becoming for a woman to
show before others too much happiness at her husband's
coming—ran out, into the woods. Only then did we realize that during all this long and tormenting time what we
really loved in Father was unattainable happiness. He had
been a marvelous dream. Now he was here. We loved
him, but of the dream and happiness there was no more.

Father took joy in his sons and his house. But it was a
worried joy. Had internment weakened and broken him?
What happened to Father happened, in fact, to the majority
of Montenegrin officers. He was an opponent of the unification, though he was hardly a fiery supporter of King
Nikola. In the modern Serbian army he, a half-peasant,
could not advance. He felt that his homeland was also
thereby belittled. The sacred things of his youth were
insulted—the Montenegrin past and name and arms. Some
Serbians called the Montenegrins traitors and threw up
into their faces that they, the Serbians, had liberated them.
The Serbians sang mocking songs, one about how the wives
of each greeted the Austrians—the Serbian women with
bombs, and the Montenegrin women with breasts. All this
gave offense and caused confusion. Those around King
Nikola acted dishonorably in the time of tribulation, but
the ordinary soldiers felt no guilt or shame. They died,
suffered, and languished in prison camps.

Dissatisfied with the new state of affairs, Father nevertheless accepted service as gendarmery commandant in
Kolašin. It did not help to think much about it; he found
himself in the tormenting position of having to act against
his thoughts and desires.

The course of history was changing, and one could not manage to warm himself at two fires at once. Choosing between conviction and a better life, most, including Father, decided in favor of the latter. Must it be so? Is this not a deliberate rejection of something that is peculiar to man alone, free thought, that which is most human in man?

The winter was on the wane, and spring had not yet begun. Everything was infused with a chill damp. The snow had turned to slush, and there was no dry land.

Late one afternoon, in the middle of the village, a column of horses and men came tramping through. There were many of them, two or three hundred. Among them were some men in half-uniform—our men, Montenegrins. They were leading mules loaded with ammunition and cannon and machine-guns. I stood in front of the village inn while the cavalcade moved on and on. Never before had I seen so many men at one time at such close range.

Five or six peasants in front of the inn stared darkly as the army passed by sullen and silent. The peasants took off their caps when a knot of horsemen appeared. Someone mentioned the name of Boško Bošković. No one pointed him out, however, nor did anyone ask about him, as though they did not dare. There was something foreboding in all this. Nobody replied to the salutations of the peasants.

Boško Bošković was the chief of the district in Kolašin. He had been a Montenegrin officer, the commander of one of our field battalions. He was a native of Donja Polja, in our district, the son of the renowned insurgent and field captain Lazar Bošković. At that time nearly every Montenegrin whose father was at all famous derived his own last name from his father's Christian name. Most everyone, however, except for a few of the older men, called Boško by his family surname. Somehow it sounded

fuller and stranger, and there was something special about
the combination Boško Bošković. As an officer, Boško had
distinguished himself in battle with his heroism and daunt-
less spirit. He was severe, yet close to the soldiers. He was
dogged by the reputation of being a brave yet rather stupid and willful man.

One had to go far to find someone who did not know
Boško's name. It was on everyone's lips just then, wherever one turned. I, too, was already well acquainted with
it. For one thing, Father constantly talked about him,
because they were fairly good friends, though Boško was
younger by about ten years, and, besides, they shared in
the government of the district. Boško was a determined
supporter of unification with Serbia. His name was associ-
ated with a wild and unyielding spirit, and even more
with the cruel execution of tasks either assigned to him or
else which he himself chose. Just as his father, some forty
years before, had raised the Poljani * to arms against the
Turks, and later ruled them with fines and the whip, so
Boško now subjected a whole province to the new government. If there was a need to find a man who would
quickly break the resistance of the opponents of unification with Serbia, and it seems they were not in a minority,
then Boško was that man. By lineage, bravery, even appearance, he alone was able to inspire awe and respect, and
by his cruelty to strike terror into the bones of his opponents.

The opponents of unification were not united. Some
were supporters of King Nikola; they were those who
yearned for an independent Montenegro. There were,
however, others who were not satisfied with the way unification was being effected—through occupation. Then
there were simply many dissatisfied peasants, men whom
the war had uprooted from their previous lives and occu-

* Inhabitants of the plains centered on Bijelo Polje, in eastern Montenegro, on the River Lim.

pations, who looked to the others, not knowing what else to do. There were also men who were dissatisfied but did not themselves quite know why; they were caught up by the rebellious times. All of them were opponents of the new regime, which, for them, differed little from an occupation except that it was more efficient and strict, since it was managed by men of their own blood and language. Variegated as they were, the opposition comprised a majority. It was partial neither to the old regime nor to King Nikola. This was a majority of the discontented, who had no clear knowledge of what they really wanted.

Though I was not able then to make such differentiations, I was keenly aware of the dissatisfaction, and—in a special childish way—even swept up by it. There were already many rebels and, at first, they roamed through the villages almost freely. As the new government became entrenched, though, their numbers quickly fell.

Todor Dulović also rebelled. He had become famous before, and now immediately became the best known of the guerrillas, and thus the most serious enemy of the new regime. Therefore, Boško Bošković set his cap for him. Boško had other motivations as well. Todor was a nobody, an ordinary peasant who had gained distinction without regard for anyone. Renowned even during the times of the occupation as a rebel, he was now regarded as the most ardent inciter of rebellion.

Todor's reason for continued rebellion under the new regime seemed, on the surface, insignificant. In Bijelo Polje, in the middle of the market place, some Serbian noncommissioned officer took Todor's cap from his head. On this cap were the initials of the Montenegrin king, N.I., that is, Nikola I. Todor immediately tangled with the officer and dug his heel into the Serbian insigne of King Peter. When the gendarmes rushed in to arrest Todor, he escaped.

Todor had two grown brothers, younger than he. Petar, the first after him, was a quiet and unassuming lad whose only concern was his work around the house. He was pale, slender, with soft lines to his face. Not even during the occupation did he want to go to the forest; he merely avoided the enemy. Mihailo, the third brother, was a youth of eighteen. He was in the high school in Kolašin at the time. Blond and lanky, with sad, shadowy eyes, he wrote love poems and paid court to spinsters and widows. Apparently nothing else interested him. Their father, Vučeta, was a quiet, simple man, a Montenegrin of the old stamp, with long mustaches and a long pipe. His speech was grave and lofty, as though he were on horseback or had captured whole cities barehanded. He was, unlike most of the backward older men, a fiery supporter of King Nikola. Too old to go into the forest, he encouraged Todor, his eldest son. This was a point of honor with fathers in Montenegro, as though they would love their sons even more if they defended Montenegrin independence. Old as he was, Vučeta was frequently jailed and beaten on account of his son. He bore it all calmly, with pride and even delight, undaunted, and convinced, if not in victory, then in the justice of his and his son's cause.

In other regions there were rebels even more famous than Todor—Žvicer, Bašović, and others. But Todor was distinguished among many, and perhaps above all, because he was never cruel or inhuman, though he had killed many men, spies and gendarmes. He never went berserk, even when slaying opponents. In Todor can best be seen what the others were like and what became of them. Todor was not the reason that the army passed through our village. But because of that bloody march he found it impossible ever again to return from the forest.

Largely under the influence of Communist propaganda, the Poljani refused to pay taxes and disclaimed any obedience to the government. This propaganda was carried out mainly by two teachers—Milovan Andjelić and Pavle Žižić. Žižić had been elected a Communist national deputy.

Even my grandfather, a man of traditional outlook and the most prosperous man in the village, joined the rest in following him, thinking that it must be all right, if one did not have to pay taxes.

Boško was particularly infuriated by the Poljani because they were his own clansmen and had been soldiers in his battalion during the war. His unrestrained fury quickly spread among his troops, especially his entourage of Montenegrins. He declared: "I know what these Poljani need—let them feel the whip, as my late father had them do."

One could tell just from the peasants in front of the inn how unpleasant the arrival of these troops was to them. They kept their mouths shut, not knowing themselves what would happen. The troops were passing beyond my sight, down below the graveyard when suddenly a volley of shots was heard, dull and soft against the damp snow, an ominous sound in the stillness of the frightened village.

Quite by chance, Todor's own brother Petar had encountered the troops as he rode along. He got off the road, but somebody in Boško's entourage recognized him and maliciously pointed out that he was Todor's brother. Boško ordered a gendarme, a Serbian veteran of the Salonica front, to fetch Petar. The gendarme, on a tall and swift army horse, and Petar, on his slight mountain pony, began to race across the field through the thick snow. Petar knew that Boško, riled up as he was, would have him whipped. To be beaten without taking revenge was a great shame, yet how could he, a peaceable peasant, avenge himself on the powerful government? So he fled, but he could not get away. The gendarme was in no mood to chase after him. He ordered him three times to stop, according to regulations. Then he fired his French rifle, toppled Petar from the pony, and left him there in the middle of the snow-covered field. Petar gathered up his spilled guts, but could not quite reach the nearest peasant house. A peasant woman dragged him into the house—the others did not dare—and he died that

same night, in great torment. Two days later he was buried
in our graveyard. None of the peasants, except for a few
women, went to lay him to rest, so great was the fear.

Boško continued on to Polja as though nothing had happened. That night he arrested a large number of Poljani,
among them the schoolteacher Andjelić. He needed neither
an informer nor a guide; he knew all the Poljani to a man.
He picked about forty of the arrested men. It was said
that there was not one among them who had been a good
soldier. He had this group taken separately to the schoolhouse and beat them, one by one, with his own fists and
boots, cuffing them and pulling their noses and whiskers.
Blood was smeared on the school benches and walls. The
whole clan had been humbled, crushed. Boško then called
a meeting of the clan, and established order and obedience.

Everyone knew the day Boško would go to Kolašin. And
everybody expected Todor to attack him on the way and
to avenge his brother, and the clan. Boško, however, paid
no attention to this. He set out in broad daylight, not hiding from anyone, though he did, to be sure, take all the
precaution of numbers. He was attacked at Mali Prepran,
opposite our house. The guerrillas could not get close, and
fired from a distance. Boško did not even get down from
his horse. His troops chased the frozen and hungry guerrillas into the woods and the deep snow before they knew
what had happened. The Polja rebellion had ended without much bloodshed, though with much humiliation for the
rebels.

The rebellion at Rovči was an entirely different matter.
Here the resistance was on a greater scale and stronger.
The Rovči clan lived in secluded valleys, and never in the
memory of man had they been humbled by anyone. The
Austrian army had hardly touched there. Among the Rov-
čani, who were accustomed to freedom, the supporters of
King Nikola and Montenegrin independence predomi-
nated. They set up their own government, which did not

recognize the new authorities in the district. Not looking for trouble, my father entered into negotiations with the Rovčani and agreed not to send gendarmes into their district if they behaved themselves and did not attack the authorities. Neither side, however, held to the agreement. A rather large patrol of gendarmes went to Rovči and was attacked. Some five or six of them were killed, the rest disarmed and stripped. This led to the rebellion and the march on Rovči.

From all sides came the army, the gendarmery, and the national guard to strike at the small and unsubmissive clan, shut up in its fastnesses. It found itself isolated, unarmed, and disunited before a force it had never encountered in the past. The artillery crashed down the peasant chimneys as though at target practice and frightened off people and cattle into the woods and ravines. Except for a few, the Rovčani surrendered to force. All resistance was broken, and, indeed, there had hardly been any.

The Rovčani were treated with cruelty and insult. Their houses were burned down; they were pillaged and beaten. The women had cats sewn in their skirts and the cats were beaten with rods. The soldiers mounted astride the backs of old men and forced them to carry them across the stream. They attacked the girls. Property and honor and the past— all this was trampled upon.

Not even my father could get out of taking part in the march on Rovči. He himself was not known for cruelty, but his gendarmes went wild. His conscience always hurt him because of the Rovči campaign, and he would never speak of it afterward. He could find no serious justification for his action. Having accepted a job and its duties, he had been led on despite his will—the fate of all those who have renounced their convictions for the sake of the necessities of life without becoming convinced that their convictions are wrong.

Yet there was something Father used as a kind of justification. Those brigands who had stolen our bull during the occupation had been from Rovči. Standing next to the cannon that was trained on Rovči, Father would cry out with each volley: "Ha, my bully Spot!" Like others, he could not help but enjoy as revenge an attack he was carrying out as his official duty. It was precisely this enjoyment that the people could never forgive. This gnawed at the conscience of us children for years.

Conflict, hatred, and mistreatment, killing and looting did not end with the Rovči rebellion. They continued, as resistance by the masses spread on all sides in the form of small fighting groups, and in other ways as well.

It was at that time that some peasant climbed up Bablja Greda, a peak overlooking the bridge on the Tara that leads to Kolašin, and carved into the cliff words that glorified King Nikola and Montenegro. Everyone who crossed the bridge would have to see the white letters on the yellow rock. Since no one else could climb up to blot out the sign, it stayed there. This was an act of great boldness and impertinence, in defiance of both human and natural law, at a time when all resistance had already been crushed, except among a few rare individuals who continued to hide in the crowd or in the woods. It was as though that peasant wished to say: We were here and fought and believed in the independence of our country, and let this be known, whether it be foolish or wise.

The Rovči rebellion and the struggle with the guerrillas brought dissension and trouble into our house. We, too, became divided, like the Whites, the supporters of unification, and the Greens, the supporters of Montenegrin independence and King Nikola.

The reasons for our dissension, however, were different. Father's own disquiet infected us as well.

Though he may not have known for certain, Father could have guessed that the guerrillas, especially Todor and his band, were coming to our house despite the fact that Father was the commandant of the district gendarmery.

Somebody must have said something about it to Father. He hurried back home on horseback all alone, from Kolašin, arriving in the middle of the night. He immediately confronted Mother with the charge that she was in league with the rebels, all the while shouting and stomping about in his boots. Having gone back on his own convictions, with the blood of rebels already on his hands, Father now carried things to their extreme. Enraged, he drove Mother out of the house. We children joined her and found ourselves in the woods, out in the snow, wailing and in tears. After Father left, we all went back to the house as though nothing had happened. That is how it goes in quarrels between husbands and wives. The guerrillas continued to come, but more rarely and with greater secrecy. Now we kept this from Father, by tacit agreement.

Father did not convince even his own family, let alone his superiors and colleagues, that he had sincerely adopted the convictions his job imposed on him. He was transferred, then placed in the regular army, and finally pensioned. The same things happened to other Montenegrin officers. Their efforts on behalf of the new regime were rewarded, but they themselves had become superfluous. Like the others, Father returned all the more dissatisfied, not knowing what to do with his unexpended energy, as though he was no longer able to adjust to life.

So, too, old Montenegro was all out of joint. Her mountains and crags still stood, but she herself had fallen, sunken in hatred and blood, seeking but unable to find herself.

Though death was everywhere on a rampage, both death and birth were still unknown to us children, for they were something outside the house.

Then there came a new life; my sister was born in the summer.

Like the other peasant women, Mother had been working all day. That evening she hurried off to milk the cows and to put the children to bed. It was a warm summer evening, so she then went to the hut in the pasture—she was ashamed because of the children—and gave birth. In the morning we found her worn out by the fireplace next to a cradle in which something was squirming.

As it happens in life, in the footsteps of birth came death.

Life about us seethed with violence and cruelty, as it was. Death, including violent death, was a common, though always disturbing, event. At home, however, there had been peace and happiness, until death stalked in to take Grandmother Novka and Uncle Mirko, one after the other, planting in the middle of my childhood two sorrows like two black marble gravestones.

Grandmother Novka lived ninety-three years, if her memory was right—and it certainly was—that she was forty when the famed Turkish hero Smaïl-Aga Čengić lost his life in Drobnjaci on Mljetičak Plain. Still, we children found it incomprehensible that she should have to die.

Living amid evil, she recalled both happy events and

great men. She remembered Bishop Njegoš well. She saw
him at a public gathering at some church in Pješivci. His
head was above all the people, he was so tall. She spoke,
too, of his unusual beauty, the ornament of all Montenegro.
She also remembered Smaïl-Aga, though not as well. He
was broad-shouldered, strong, had a large head, and was
already well along in his years when he lost his life. He was
severe, but one of the more just agas. In fact, he himself
would not have had such a bad name with the rayah had it
not been for his son Rustem, a bully from whom no peasant
girl was safe.

But what evil things she remembered! How many murders, massacres, decapitations, plundering raids, and smoldering ruins! How could her heart endure all this for so
many years without withering? It was little wonder that
she was such a shrewish woman. But she managed to survive all evils, to raise a brood, and to look upon their
progeny.

She was one of those Montenegrin women whom no
calamity or catastrophe could keep from fulfilling the purpose her life was meant to fulfill—to breed male heirs and
to keep from ruin the house into which she had come.
There was something not only traditional but inborn about
this. Without such women, and they were all like that, this
people would not even exist.

Grandmother was still in full possession of all her senses,
though she was shriveled. Just, fearless, and resolute in
everything, she did not permit old age to direct her thoughts
to the other world, but kept them wrestling with this
world, of whose afflictions she never complained. She did
not pray much to God, just enough to preserve her from
phantoms and evil spirits. It would be hard to define in
what exactly she did believe, but certainly not in a Christian
deity. Rather, she believed in all kinds of apparitions and
phantasmagoria of ancient pre-Christian times. She never
swore, but in wrath she pronounced maledictions, short

dire curses, which struck terror and chilled the blood. These curses were terrible for the pictures they evoked and not for the way in which she said them, for they did not come from her heart. They slid from her tongue, in the passion of anger, which she soon forgot. Her fits of wrath were sharp and sudden, but her loves were deep, though she was loath to show them, hoarding them instead.

One could not really say which one of us three grandsons she loved most. It was I who most frequently found in her a defender against Mother and my brothers. For this, to be sure, I loved her more than my brothers did. She always knew how to wheedle me into eating. She was easy to fool, too. By stubbornly refusing to eat I could always get the tastier morsels. One had to know just how far one could go. Like all children, I knew this well. Beaten, in tears, I would fall asleep in the solace of her lap, against her withered, bony bosom.

Christmas was a holiday that inspired joy long in advance. On the day before Christmas, even before dawn, the Yule log is sought in the forest and preparations are made for the holiday festivities. This particular day before Christmas was just like that. But Christmas dawned in sorrow. Grandmother was on her deathbed. She had been ailing for some time, and yet it all happened so unexpectedly. She had seen everything there is to see in life, yet she longed for life, and the family wept with her.

The death rattle came suddenly, at dawn. Even in her sleep Grandmother began to breathe heavily and in short, hoarse gasps. Then she grew stiff and could no longer speak. In the morning the peasants came, and only one of her sons, Uncle Mirko. Everybody said that she was dying, so they lighted a candle by her head for her soul—unconcerned that she, too, would see that she was dying. No one drove the children from the deathbed. Somebody asked if she were conscious and could recognize people. Uncle Mirko rose and began slowly to pace up and down the

room in front of Father's bed, the one on which she was
lying. She had loved Mirko the best. Her bulging and, till
now, frozen eyes slowly followed after her son, as if to
leap after him, and tears flowed from them. The son knelt,
crushed, by her side and placed his large head on her
breasts. She did not have the strength to embrace him. All
she could give of her love for him came through her eyes,
in her last farewell to the world, and to life, as though it
was all the harder for her at that moment to leave because
her soul had abided in them so long.

Grandmother gasped twice, twitching feebly, and then
the rasping ceased. Uncle said, weeping, "You will not wait
long for me, Mother." A wailing ensued in the house.
Sobbing, my brother reproached me for not crying over
Grandmother. She had died, but I did not see that she had
given up her soul, which I imagined to be like a tiny wisp
of mist floating upward into the sky. Grandmother's death
was unforgettable and shocking because of the simplicity
with which a loved one had become transformed into a
thing and actually ceased to exist.

I kept Grandmother within me long after.

Whenever Mother beat me, I would say, "Grandmother
hears it all, just so you know, and she will curse you." But
just how could she hear, and exactly where was Grandmother staying? Somewhere in the ground and, at the same
time, in the sky? At night sometimes I could not fall asleep
for a long time thinking about her and remembering her
favors and, most of all, her death. I was never afraid of her
ghost, that is, if she ever came to my mind or appeared to
me. A severe and rough woman, with a will of her own,
she was for me a kind old lady who protected me to her
last breath from everyone and everything and who told me
spine-tingling stories.

Grandmother's death seemed to goad Uncle Mirko into
taking from life what was good for the taking, with all the
last strength left in him, without regard for anything, even

his good name. Always a man of his word, sober and upright, he now grew wild and berserk as never before.
Though in his sixtieth year, he decided he must have male
progeny. In Montenegro anyone without male children
was cursed. With him the longing for a son became a
gnawing disease and, moreover, an unavenged wound. He
did everything, visited fortunetellers and drank potions. His
wife had long ago stopped bearing children, some thirty
years. They had once had a son, but he had died. He spoke
of him like one possessed, of his intelligence and beauty,
though the child had died while still in the cradle. About
his wife he said crudely, "I planted seeds in her for forty
years, and all in vain. No seed can sprout on rock, and
out of her you couldn't get even a stone." He counted his
daughters for nothing at all.

Realizing the unrelenting force of his yearning, and feeling herself to blame, his wife declared, "Let him bring another younger woman beside me. I will be like a mother to
her, but I have spent my whole life with him, and let him
not drive me away." He paid no attention to this but
chased her away, as though she were a leper. He brought
into his cottage a widow with two children, all in the hopes
of her giving birth to a son. She, in turn, settled easily on
the property, but she did not last long. Either he himself
became convinced that it was all too late and in vain or
else he grew angry for some other reason. At any rate, out
she went, too.

At odds with his own daughters, he had no one any more
to care for him, and, brought low by ill health and misfortune, the spring after Grandmother's death Uncle
Mirko came to stay with us. In addition to his unhealed
wound, he had developed consumption in his old age. He
was a strong man and took long in dying, the whole spring
and summer. Of all us children he liked me the best. In that
love there was something unwholesome, which affected
me, too. Without male children, like a dead log in a forest,

in the last months of his life he transferred all his yearning and hopes, property and favors to his nephew. He never for a moment let me out of his sight, as though he wished to borrow life from me to preserve his own life, which was so obviously waning.

The longer he ailed, the thinner he became. His head seemed to grow bigger and more pronounced, his eyes burned with a furious green brilliance, and his speech became more terse. His teeth, healthy and handsome, seemed larger and whiter, and his gums hard and brittle. He became more and more yellow, though he smoked less and less. From his wound there oozed a constant bloody pus, which, on a summer day, gave off a sour smell of putrefaction. He coughed constantly and spat thick globs into a pot. To get me to stay with him all day long, he invented the excuse that I was needed to protect him from the flies. And this I did, day after day.

Outside there was spring, sunshine, swift streams, and blue whirlpools, and here—but I stuck it out somehow. How could I leave him? How could I help liking a man, a sick man at that, when he loved me so much? The more I suffered at his side, day in and day out, the more it seemed that I became bound to him. Mostly he just kept his eyes shut. Yet he was always awake, ready to say something unexpected and to anticipate by some word someone else's thought. He would talk, too. "It's tough on you," he would say. "A nice day, but wait just a little while longer. I hate to be alone. I have no one but you." Or: "Tell me a poem to while away the weariness, yours and mine." I was already old enough to read the folk epics, and so I recited to him.

Early one evening, however, a sudden turbulence seized him, as though he had at last understood what he had been turning over in his mind all those wild days and nights. He clutched at me with both fists, like steel traps, and brought me close to his eyes, which smoldered with a green phosphorescence deep in his skull. He declared: "Remember, let it be your sacred task—to avenge me. You shall avenge me! Get an education, but avenge me as best you can. Only remember—avenge me, so that the earth will rest lighter on me." I promised I would and, torn asunder, I sobbed, "No, Uncle dear, you shall not die, and I will avenge you, I will, I will. . . ."

Revenge is an overpowering and consuming fire. It flares up and burns away every other thought and emotion. Only it remains, over and above everything else.

The word "blood" meant something different in the language I learned in childhood from what it means today, especially the blood of one's clan and tribe. It meant the life we lived, a life that flowed together from generations of forebears who still lived in the tales handed down. Their blood coursed in all the members of the clan, and in us, too. Now someone had spilled that eternal blood, and it had to be avenged if we wished to escape the curse of all those in whom the blood once flowed, if we wished to keep from drowning in shame before the other clans. Such a yearning has no limits in space, no end in time.

When he had returned from internment, my father was nearly killed by one of the descendants of Captain Akica Ćorović, who had been killed fifty years before by my father's father. The Ćorović clan hankered to kill my father, and his very rank encouraged them. They egged one another on—a captain for a captain! They had a good memory for blood. The Djilasi were hardly forgetful either.

I had a close friend in the seventh grade, the kin of those who had killed my grandfather, though I knew nothing of this kinship. I brought him home one night around Christmas time, so that he would not have to go home on a lonely road in the dark. When my father learned who he was, he forbade me to sleep with him because, as he said later, he did not want anyone with our blood on his hands to breathe into the soul of his son. Father did not sleep a wink that night, but tossed and turned on the bed. What were his thoughts? What was he weighing in his mind? The next day I escorted my friend home, and only then did my father tell us that a male child of an enemy clan had spent the night in our house. After the recess I met my friend again, and he, too, had obviously learned the awful secret from his people. We avoided one another and our friendship was smothered, without either of us admitting the real reason, for our still childish breasts were straining for revenge.

Vengeance—this is a breath of life one shares from the cradle with one's fellow clansmen, in both good fortune and bad, vengeance from eternity. Vengeance was the debt we paid for the love and sacrifice our forebears and fellow clansmen bore for us. It was the defense of our honor and good name, and the guarantee of our maidens. It was our pride before others; our blood was not water that anyone could spill. It was, moreover, our pastures and springs—more beautiful than anyone else's—our family feasts and births. It was the glow in our eyes, the flame in our cheeks, the pounding in our temples, the word that turned to stone in our throats on our hearing that our blood had been shed. It was the sacred task transmitted in the hour of death to those who had just been conceived in our blood. It was centuries of manly pride and heroism, survival, a mother's milk and a sister's vow, bereaved parents and children in black, joy and songs turned into silence and wailing. It was all, all.

It was our clan, and Uncle Mirko—his love and suffering and the years of unfulfilled desire for revenge and for life. It was his death. Vengeance is not hatred, but the wildest and sweetest kind of drunkenness, both for those who must wreak vengeance and for those who wish to be avenged. Uncle's last hours were filled with a yearning for revenge, for male progeny. Were they made happier by my vow?

There was much wailing and keening for Uncle, as

though the cliffs and crags had toppled down. As they carried him away, along the ravine by the house, a woman uttered a long-drawn-out howl. The coffin led the procession. I felt then, for the first time, a numbness and a change take place in my body; I was completely dazed and shocked, without knowing myself why. This feeling seizes me whenever death touches me, whenever it becomes apparent and irrevocable. This, then, was death. Uncle's coffin made its way into the woods, and became forever lost from sight. Uncle had died. Grandmother had withered and faded away, but Uncle died—while still strong, though gray and spare.

His clothes were kept in the house. According to custom, they were placed in the middle of the room, as when they had been worn by a living man—a cap, and underneath that his costume, with sleeves folded and weapons in the sash. The clothes looked as though their owner had just gone, never to return again, and thus they served to spread an even more inconsolable sorrow and empty ache than the dead man himself. Uncle was a handsome dresser, and his gay clothes made his death all the more undeniable and palpable.

From life, and from my life, two lives were uprooted. Life and my life thereby remained with wounds that would never heal. A loved one may be forgotten, but the emptiness that he leaves in us stays on. We live with our own lives and also through the lives of others. So it is that we die, too.

Even in a rich country in normal times, the life of a peasant child is a hard one, spent in bitterness and even in peril. He frequently falls ill when there is none to heal him. He is constantly beaten, by everyone, even at school, regardless of what kind of pupil he may be. Whenever I was switched across the hands, I observed how bony my fingers were and how lean my palms; my fists were like little birds.

The whole earth is engaged in a constant struggle, and the child is dragged into it from the time he first becomes conscious of himself. Men fight with one another, and the animals prey on each other. Poisonous snakes lie in wait behind every rock, behind every tuft of grass. The earth is sown with thorns and rocks.

With means for destruction such as he had never known before, and unleashed by wars and rebellions, contemporary man has laid waste to everything around him. He has extirpated wild beasts, devastated forests, and destroyed fish, the trout and salmon of the clear swift brooks. From our earliest years we knew how to handle weapons and to hunt animals and fish with our rifles. The older boys knew how to set off grenades and kill fish with them in the pools. We younger boys loved the game of depositing a grenade in the hollow of a tree and lighting a fire beneath it, watching from our shelter as the explosion blew to bits the century-old trunk.

While he was playing with pistols, my brother once wounded me, but Grandfather made me well. He was known throughout the countryside as an expert in healing wounds and making balms. The peasant child grows in constant struggle and pain with wounds, sores, lice, bare feet, hunger, and neglect. He has to watch the cows and help his elders in their work, though he be sleepy and tired. If it had been a time when men were gentler and more at peace—but, being what it was, my childhood held very few happy memories. There was only childhood itself, full of a luster of its own, of a growing recognition of the world and its games.

Aimless wandering through the mountains remains to me a memory of unspoiled beauty. The mountain draws a man to itself, to the sky, to man. There the struggle that reigns within everything and among all things is even more marked, but purer, unsullied by daily cares and wants. It is the struggle between light and darkness. Only there on the mountain are the nights so vast, so dark, and the mornings so gleaming. There is a struggle within everything and among all things. But above it there is a heavenly peace, something harmonious and immovable. The heavens impose the question: Who are we? From whence have we come? Where do we go? Where are the beginnings in time and space? No need to feel impatience or anger over the answer, no matter what it will be. Men on the mountain are an even greater mystery. And the stars are as near and familiar as men. The earth and sky and life become unfathomable, daily riddles that arise spontaneously, and that demand an answer. And so, forever, all must give reply. All, from the old man to the child. For the mountain is not for a tale, but a poem and for contemplation, and for purified emotion and naked passion. Life on the mountain is not easier or more comfortable, but it is loftier in everything. There are no barriers between man and the sky. Only the birds and the clouds soar by.

A summer outing in the mountains is for the young people, who yearn for the effort of a climb, to unleash their

strength, and for the chill nights and mornings, to bathe
in freshness. On the mountain everything is rough and raw,
but clean as in a song or a maiden's embroidery. Life seems
to shift from man to nature. Even human life becomes all
enveloped by the sun, by the verdure and the blue,
drenched by them, less ashamed of its passions, less withdrawn, like a herd of horses galloping freely across endless
pastures in a time when they had not yet been subdued by
man.

One goes to the mountain also for a holiday, to rest the
body and to give free rein to the mind, to play and thus to
melt into nature and the universe. The beauty of the mountain is not merely in the clean air and diamond-cold water,
which cleanse the body within. Nor is it in the easy life.
Its beauty lies in that ceaseless and all-pervading effort and
exertion, which are not really oppressive. Stern in appearance, the cleanliness of its waters and air overpowering and
yet invigorating, the mountain nevertheless dances in luster
and color, and forces all creatures, above all man, into
dances of spirit and body that are guileless with all their
boldness and abandon.

On the mountain there is something for everyone—for
the young, brightness and play; for their elders, sternness
and constraint. Sorrows are more sorrowful there, and joys
more joyous, thoughts are deeper, and follies more innocent. The cattle immediately come to life there, as though
fattened by the freshness; they become playful in the fence-
less spaces. Like a river or a city, each mountain has its own
life and own beauty. Mount Bjelašica was special because
its streams and grass reached to its uppermost heights of
wide and rolling meadows. She was warm in her coldness
and gentle in her steepness. Her air was as chill as on the
heights of a glacier, yet the sun shone as hot there as on the
villages in the valleys. In her pastures one found a bower
and a haven.

Every clear evening in the middle of the pasture a huge

fire is lighted; around it dancing and singing surge up. The
fire is not lighted because of the cold, but to radiate joy
and light, to enliven the mountainsides and peaks, and to
join the youths and maidens in their mad gay dance. In
these camp meetings, in their dances and songs, there is
something irrepressibly savage, something just barely and
invisibly kept from tearing loose from human bonds and
from reverting to a primeval wanton and joyous madness
such as man had never known. When the fire and the
dancing subside, shrieking and laughter break out on all
sides, and then begins a wild chase and commotion. Impassioned youths dart after the maidens, pinch and embrace them. The maidens grow even more resilient and
elusive than in the dance, as though the darkness has jerked
them up short into a life of strict rules which decree that
they can dance and joke in public but must be virtuous and
unapproachable in private. The widows, who sit before
their huts listening all aquiver to the dance and who lose
themselves gazing at the frenzied motions of the shadows
against an endless sky, creep into their beds beside their
children, crushed by an onslaught of emptiness and bitterness. All seems to die, in a twinkling, but constrained hearts
still beat loud on the hard bedding, bright and sinless
thoughts sprout and spread, while murky desires burn out
and smother one another. A little longer—and then the
morn. The first cock brings peace and the dawn and daily
cares and tasks.

Even without the mountain the village boys and girls
learned much about love life from coarse and unabashed
jokes or by watching the pairing of animals, especially
cattle. Whenever cows were coupled, the girls or young
women who brought them retired, while the lads made a
point of bringing the bulls while the girls were still there,
making rude jests the whole time. There were games that
were even ruder, such as jumping on little girls and on
heifers, games of which the boys and girls were themselves

later ashamed. The children began to play them while yet quite young, but these were games. The mountain, however, seemed to evoke in children passions that were much later to flare up.

I was ten years old. I had already completed elementary school and was preparing, inside myself, to go to the city. It was as though I had reached an understanding with my childhood to end it there in joyous exertion. That summer I spent in the mountain, with the cattle. I had to get up early to drive the animals to pasture. The mornings were oppressively bright, but the fresh heat and quiet of the day was welcome, and so was the deep slumber and oblivion of the night.

Kosa was a hired hand, a strong and sturdy mountain girl with a rough face but gentle yellow eyes. She was good-natured, gay, and tireless in her antics with the boys at camp meetings. I watched over the cattle, while she did everything else. She was one of those busy bees who managed to do everything and yet have a good time. We slept together, in a cramped lean-to next to my aunt's large hut. Even before then I liked Kosa, who was always gay and good at everything. But it was on the mountain, that summer, that I fell in love with the enticing warmth and softness of her body. Each time she would return from the campfire, still in a sweat, she would lie down beside me, nestle up against me, and place my hand in her bosom, glad that she could uncover herself next to a boy. The moonlight cut through the beams like flashing swords. I lay there aroused, unable to fall asleep again. I felt a secret delight—I did not myself know why—spring from Kosa's body, which now seemed like a part of my own. Never before had I felt such a sensation. But, neither the mountain nor Kosa helped. With that delight and knowledge of her body came also shame, and so, roused out of my sleep, I lay there taut and motionless. Yes, Kosa was impassioned; she was embracing the boy and getting the boy, instead of

some older fellow, to caress her. Or maybe she was hugging
me as she would a younger brother? And then again
maybe . . . ? She, too, was awake and in motionless silence. And I could not, dared not, do anything but tremble
inside and quiver, powerless to solve the riddle of her body
and her desires.

Everything else, as well, is revealed on the mountain and
becomes simpler and clearer.

Down below, in the villages, tribal and clan divisions
were already beginning to fade. The mountain, however,
had been divided from earliest times. It was known to
whom every peak and spring belonged, as well as the pastures and meadows. The tribes no longer fought over their
valleys, but the shepherds still fought over their grazing
lands, made up mocking jests and howled derisive songs at
other camp settlements.

Just as every family in the village was proud of something, so every camp in the mountains was proud if its
bull lorded it over the others, or if its horses were swifter,
or if its lads could heave rocks farther or outleap the lads
of other clans. Good householders that they were, my
uncles valued their good cattle as much as their own good
name. Not only were they rich, but all their animals were
distinguished for their size and fatness. My uncles insisted
most of all, of course, on having the strongest bulls and
oxen. They were the kind of men who would buy a bull
just to have the strongest. Spendthrifts in everything, in
this matter their caprice and passion knew no bounds.
Their bull Rusty was the lord of the camp. He had van-
quished the bulls of most all the other camps, and there
were hopes that he might become the king of the mountain. He bristled with pent-up power and was a terror to
man and beast. There was, however, on the same mountain
a bull from another camp who had not been beaten in
three years and who was now reputed to be the invincible
champion. He was a strange bull, a bit taller than the rest,

but without that thick bull neck which dominates the rest of the body. His name was Spotty. Unlike the rest he also had a surname, his owner's—Bekić. That is what everyone called him—Spotty Bekić. It had a ring to it, like the name of a prize fighter.

Bulls generally live with their own herds, except during the time of mating, when they are kept apart until the cows are brought to them. Spotty Bekić, though, lived alone, wandering over the mountain, from camp to camp, from herd to herd, fearing neither wolf nor brigand. The Bekići were not rich, but their bull had made their name famous over the entire mountain.

For two years, ever since Rusty had become number one in his camp, the shepherds looked for a fight between him and Spotty. But an opportunity never seemed to come. Besides, my uncles avoided such a fight for fear that the Bekići might encroach on their fame and glory. However, Spotty found his own opportunity. One summer evening, above our camp, there arose a piercing bellow. That was Spotty, who was already known to many shepherds by his voice. He came straight at our huts, as though he knew where his opponent was, and began to nuzzle our cows. Here was an opportunity for a duel. Half the camp gathered around in a trice. A bull is more belligerent when he is defending his own herd. If he was ever to win, Rusty had to do so now—on his own home ground and with his own cows.

We brought them together. Rusty went into battle somewhat reluctantly. We could tell from the very first clash that Spotty would win. He was swift and skillful. Rusty could not make him move. Then, as if by agreement, my uncles' shepherds began to beat Spotty with anything they could find, just to help Rusty. Spotty endured this for a while, wheeling around and around, and finally he went away. Rusty chased after him rather fearfully. Spotty easily made his way up the hill. Night had already fallen. As though he was turning something over in his mind, the bull kept silent. Then, out of the darkness above, his bellow was heard once again, insulted and sad.

The next day the news spread that Rusty had beaten Spotty. A great injustice had been done to a hero. But the truth will out, even among cattle. The shepherds in Spotty's camp refused to believe the news. They chased after us for days to bring the bulls together again, in a pasture. Finally they caught up with us. Because we did not want a fight, they held our arms while the bulls were driven out into the open. The duel was an honorable one, and Rusty was quickly defeated. Spotty Bekić, that strange knight of the mountain, regained his stolen glory.

On the mountain one also felt still more the difference between the poor and the well to do. The poorest of the poor had no cattle whatever of their own, but hired themselves out to herd the cattle of others as sharecroppers. The huts of the poor were small, and their cattle always nestled against other people's cattle. They were thinner and weaker, as though they knew whose they were. Among the poor even heroism counts for less. They did justice to a heroic bull, but a man they were apt to forget.

In the village there lived a certain Sava Pejović in abject poverty, though he had once been a hero. He never had reason to go into the mountain. His children, thin and sickly, languished in the village, looking sadly out of the corners of their eyes in the hope that someone would hand them a piece of bread. It was this Sava who had killed the renowned Turk Zeko Lalević, while our village had still been under the Turks, thereby setting off a rebellion in the whole region. Zeko's house, a handsome wooden building in the Turkish style, fell to the chief officer and not to Sava. Now nobody even believed that this spare, beaten, and timorous man, who could barely stand on his own two feet and whose pipe hardly contained even the smell of tobacco, had done that heroic deed. To many it seemed that he had done the deed by accident or out of necessity, as though such heroism hardly befitted him or any of his kind. His bulls did not bellow in the mountain, his dogs did not bark, nor did his bellwether rams sound their bells. Whenever he went to the mill or into the woods, he had no pack horse but himself. He wasted away, oppressed by misery and oblivion. His heroism was a thing of the past, but his poverty went on and on.

The mountain is not kind to such poor heroes, only to those who are strong in everything. She gives, but she also takes. Only those who are strong in everything can survive in her and even grow stronger. One always longs sadly for the mountain, for its strength and purity, for the endless beauty of its peaks, whose colors blend and die in one another until all sinks in a bluish mist. The mountain has aroused new perception and feeling, though perhaps she did not do this herself. As everything within man is first expressed as an experience and a picture of the world, so such emotions are bound with the mountain and remain in us, in me.

PART TWO

The Men and the Times

I had already heard about both the city and the high school. As with everything else, the real thing was not as it was pictured.

In the fall of 1921 my father boarded my brother and me with our cousin Draguna of the Ćetković family. They lived in Bakovići, nearly an hour's distance from the high school in Kolašin. Though Bakovići was a village, its way of life and customs were unlike those I had known. It was near a town which was small but nevertheless a town.

Draguna's family was as different from ours as their village was. They were peasants like us, yet even poorer. This did not prevent their regarding us as commoner. And with some justice. With them everything was cleaner; their house was tidier, and their courtyard was always swept clean. With us everything reeked of cow manure; our entire existence was bound up with our cattle. With them everything was lean and clean. They knew where every penny went. With us everything was greasy and grimy. We lived as though we had no sense of measure or value.

There were two main reasons for this difference: the influence of the nearby town and an educated son. Aunt Draguna herself was a very bright woman. She knew her way around and kept up with the times, though she was always angry at their changing fashions. Here one could see how much the way of life had changed after the wars of 1875–1878. Before, there were no such houses, nor such towns as this, which from its founding radiated order and cleanliness.

Aunt Draguna was one of those severe, sharp-tongued, and rather self-willed women. The whole village was afraid of her small green eyes and cutting words. She would scold and threaten, but without hatred or malice. She neither gossiped nor swore, but she let no one get away with anything, least of all her husband. He was a man of bygone times, simple and slow and taciturn, and endlessly smoked wild tobacco in a long Turkish pipe. He was always barefoot, as though he delighted in this. On the rare and brief occasions when he flared up in anger, he could be dangerous. In those moments Aunt Draguna would withdraw; apart from such times, she was the boss of the house. Both she and her husband were convinced of her wisdom and capability, and both submitted to them. She married at a time when girls did not choose their husbands, otherwise she would have taken another. But she was so full of good sense that she made a good marriage anyway. She made light of her husband's simplicity, but never in a way that humiliated him before others.

Draguna was of our Djilas clan. As my brother was named after our grandfather, so I was named after her father, who, like her uncle, had been killed. My father knew the story of their unhappy death well; it had much in common with our own tales as well as those of many others—ambush, decapitation, and revenge. Draguna was still in the cradle when her father died, and since she had no brothers, she transferred to my father all her love for a father she could not remember and brothers who had never been born. She did not allow even this love to enslave her. If my father did anything that was wrong or unjust, she would not pass over it in silence. Her love was the old clannish kind, but her mind was impartial.

Draguna had three daughters and a son. The eldest daughter had already come of age. The son, Ilija, the eldest child, had been made a teacher in a village more than an hour away. He was the worry and hope of the family. He was a zealous supporter of the unification and an opponent of the guerrillas and Green bands, and his family lived in constant dread, with some reason, that disaster would come to him. Ilija inherited capability and intelligence from his mother, but in him they were even sharper. His mother was haughty and aloof, and so was he; yet she was goodhearted underneath, while he had a streak of malice and anger. He was particularly irritated by the peasant ways of the household. He would not abuse his parents; his anger was shown in a special way—he would refuse to eat and would leave in a rage. This, of course, hurt them more than any abuse, though he did not wish to offend them. A cultured man does not abuse his parents, and he considered himself a cultured man.

There were three rooms in the house. One was a storeroom, one was for the family and the two of us, and in the third, the largest and brightest, slept Ilija, the only son, all by himself. So, too, everything else was divided, yet the whole family lived in dread of his outbursts. In Ilija's room there was no speck of dust or spot anywhere. It was well furnished and had pictures—not icons, but portraits of half-nude women amid red posies, as beguiling as nymphs. They were cheap and garish and offended peasant virtue, but they were respected like icons because they belonged to Ilija. Those who entered his room always left their shoes at the door to avoid disorder, as at a holy place.

For Ilija the coffee was either too bitter or too sweet. The eggs were either too hard or too soft. The bread on the table was not in a dish, as it should be. Ilija would leave the table in a rage, and his sister Petruša would weep. Everyone would scold her: "Ilija left hungry!"

It was said that his nervousness began when Principal Jojić of Berane hit him over the head with his cane. There was very little truth to this; Ilija was simply too sensitive and orderly, repelled by common peasant ways and dirtiness. He wished, at all cost, to live a cultured life, the kind he had seen and imagined.

Ilija had become a notable figure in Berane quite by

chance. Most of the students in the normal school were of age and demanded from the school authorities the right to take part in the parliamentary elections. When this was denied, they surrounded the Principal near the bridge at Haremi. He whacked Ilija, who accosted him angrily, on the head with his cane. The Principal was later removed, and Ilija very nearly expelled. The Principal's blow, which was really rather light, was transformed later, in student petitions, into a grave injury. Ilija was not given to pretending, but he acted as though the shock had deranged him. From that time on he had the reputation of slight madness.

Ilija was moderately kind to my brother and me, but we avoided him, not in fear, for he would have hit no one, but in the desire not to provoke him. Ilija respected his close and distant relatives, and, out of necessary courtesy toward his uncle, our father, Ilija, when Father visited, allowed into his room even his own father, Milosav, who remained unchanged—barefoot, dirty, and rustic. And then all the rest felt free to come in. Milosav knew, though, that he must not dare to fill his pipe with his rare green wild tobacco and light it. Father, on the other hand, acted as though Ilija's fussiness was no concern of his. He threw down his things, took off his boots, scattered ashes about, and paced the room. In vain did Petruša and Draguna clean up after him; he did not notice and behaved as though at home. We were delighted, for Ilija's severe and intolerable regimen was thus upset.

Even Ilija seemed to relax. Father was a good raconteur and was tireless in making conversation, yet this alone did not cause Ilija's respect for his uncle. There was something deeper here, which transcended his own, for us incomprehensible, life of order and cleanliness, and the disorder his uncle had brought into it.

All the girls in the village, Petruša included, yearned to look like city girls. They wore silk blouses and put on shoes before entering town. Ilija hated this artificial, impure finery, and mocked and raged about it. Milosav grumbled, and Draguna scolded in feigned anger: "See how spoiled they have gotten." They looked neither like village girls nor city girls.

Ilija raged, too, when Milosav's daughters carried eggs and milk to market. It was a gypsy business. But what were they to do? How else were they to get money for coffee, tea, kerosene? Ilija did not provide for this, and, after all, they had to care for him and keep him supplied.

Aunt Draguna realized best of all the necessity for trade and ties with the town. She spoke wisely and coolly, without regard even for her only son: "Without anyone we'll manage, without anything we won't. . . ." They say that the same words were spoken by Bishop Njegoš's mother, Ivana, just after his burial, when her husband, Toma, wondered why she was guarding the grain from being trampled by the cattle. Mothers love their sons and would give their lives for them, but nothing can hinder their comprehension of reality and the cruel necessities of human life.

The following year Ilija was transferred to Serbia. His room was kept as orderly and clean as though he were still there. It was as though the fear that he inspired reigned for all time. The room expressed both a longing and an agonizing love for an only brother and son. Ilija's lofty anger was forgotten, but the order and cleanliness remained. It seems that this is the way it has to be here; everything new must come about by making people suffer, sometimes those dearest to us.

Ilija was a stubborn and zealous supporter of Ljuba Davidović's Democratic party. Milosav was a Democrat, too, because of his son, though his son never acknowledged that his father even knew what Democrat meant; he regarded him as a secret supporter of King Nikola. A very simple and old-fashioned man, Milosav liked the patriarchal gov-

ernment of King Nikola, but as a man of the people he
also liked democracy. Ilija, a purist in everything, was a
purist in politics; therefore, he denied his father's democratic stand because it was tainted with another, an impure,
alloy.

One could not say there were many Democrats. Most
of them were the educated men. The division of the people
into supporters and opponents of unification with Serbia
faded away and disappeared. Still, it was not entirely safe
to be an open opponent; such men often hid in other
parties, ashamed of their concealment.

After the war, many peasants, made bitter by it all,
decided for Communism. But they withdrew after the first
blows the government struck, and there were few supporters of Communism after that.

Mihailo Vicković, from the neighboring village, was a
sensible young peasant who had read rather a good deal.
Blond, almost white-haired, he looked even better and
kinder than he was. He was a Communist because he loved
justice, like Christ—as Mihailo frequently emphasized. By
his gentle nature, his pleasant way with children, and because he was my godfather, he won me over to Communism
in my seventh year. Everyone laughed whenever I said
that I was a Communist—everyone except Mihailo and me.

In the affair of Boško Bošković, Mihailo was thrown in
jail and beaten. From that time on, he began to withdraw
from Communism and into himself, out of shame that he
had been beaten. Everyone noticed and laughed. He became even gentler, almost sick from goodness. I felt sorry
for him, but I did not give in as he did. Communism is
something just and for poor people, and, what was most
important of all, as a Communist I could feel interesting in
front of others. It was like hiding well in a game of blind-
man's buff and suddenly emerging in full view.

Two of my teachers were also Communists—Andjelić
and Žižić, mentioned before. Žižić was a Communist dep-

uty in parliament up to the time the Communist party
was banned. Then he abandoned Communism and devoted
himself to his teaching. Even those who were not Communists held it against him because he had given up his
ideals. Andjelić, however, persevered. He came of a well-
known and established family, was prudent and quiet, but
determined, a bachelor, and as thin as an ascetic, with the
big, blue and deep eyes of a prophet. Having once been
dismissed from a government post, he never sought one
again. He retired into loneliness on property belonging to
his brothers and slowly made something of it, though never
without a book in his hand. Andjelić's reputation grew
constantly, even among his opponents.

Then in Bakovići there suddenly appeared a real Communist, the kind who cannot conceive of renouncing his
ideals or retiring into a solitary life. He was Ilija's closest
cousin, Milovan, still a medical student. His fame had
reached us long before and had captivated us. Here was
a martyr from prison, a good student, and a simple man
besides. He would emerge from the dusk, dusty and poorly
dressed, bearded and with long hair and a red tie. One
could see immediately that, unlike Ilija, here was a man of
the people. He kissed all the grandmothers, spoke more like
a peasant than the peasants, ate with his fingers, and slurped
loudly. Ilija had always and in everything been a stranger
to us, but this fellow was our kind. We boys boasted that
we, too, were Communists. While the others laughed, he
told us seriously that every honest man must be one. Enticing words, but not quite true. There are many honest people, but few Communists.

Milovan departed as unexpectedly as he had come, as
though he were running away. Behind remained a fable
about a man hunted who escaped from every danger, never
surrendering, always succeeding.

Between Milovan and Ilija flared a conflict which had
been glowing softly for a long time. Ilija ridiculed Milovan's

simplicity as cheap deception of the peasants. He was not right. Milovan had really become a man of the people and admired their simple ways. On the other hand, he resented Ilija's gentlemanly airs, which was also not right. But the conflict between them centered on something else—their ideas. They quarreled bitterly and intolerantly, though, strangest of all, neither wished to. They came from the same clan and were nearly brothers, and members of the household tried to make peace between them, but in vain. To them, would-be peacemakers were lunatics.

There have been quarrels before between clans and tribes and between religions and nations. This, too, was such a quarrel, yet more, for it brought hatred and discord between brothers. It caused blood feuds between men of the same faith and tongue. This was something fierce and final, to the death, as it is between Montenegrins and Turks. So it must be.

I already knew some poetry, mostly folk epic. But I had not yet met any poets.

The folk epic still lives, but in the speech of men rather than as a thing in itself. The old bards, the *guslars*, were already a rarity. People still liked to listen to them, though, and I, too, enjoyed them—not so much the song itself, as the way the *guslar* told it. His sharp cries and quavers gave life and flavor to the song. Listening to his songs, I, like others, lived with the heroes and deeds of folk epic and tales. I was intoxicated most of all by the feeling that I, too, was a part of that grand narrative, which shone through the living present, the past, and the future of nations. There was something austere and exalted in the often monotonous repetition of images and phrases in the *guslar's* chant. Again and again, he depicted the trials and misfortunes through which we must live as a people, and showed us how to become men—to sing, to make merry, to keen, to create, to invent, and to produce, and, above all, to guard our honor and good name.

I had read most of the folk epics while still in elementary school. Frequently I recited them to the villagers. They liked most of all to listen to the *Mountain Wreath* by Bishop Njegoš, not only because they had heard that this was the greatest Serbian poem, but because they found in it more than anywhere else the greatest expression of their way of thinking and feeling. They found in it the essence of their ancient and still-present struggle for survival and the honor of their name on a soil that was barren in everything but men. The *Mountain Wreath* contained higher truths, their truths, truths that they had already anticipated, yet which were narrated in a more concise and lofty manner. One could stop reciting at any verse, and someone else would take it up and continue. Sometimes people would interrupt the narrator to interpret passages, ardently and long. They were not confounded even by the most philosophic passages; these they interpreted in their own way, in the light of their own image of the world and life. Many phrases and allusions that had given no little trouble to the experts were quite clear to the people because they were distilled from a life that they knew through their own experience and would have expressed if only they could. They experienced the *Mountain Wreath* as simultaneously loftier and simpler than other literature. It uncovered for them something untransitory, something that would last as long as their race and tongue survived. It was expressed in the language of every day, woven together powerfully and completely, as though it were not created at all, but existed simply of itself, like a mountain or the clear untamed gusts of wind and the sun that played on it. These people hardly knew the Bible. For them the *Mountain Wreath* might have served as such a book.

Poetry captivates and intoxicates, but poets never achieve this except when they are good men. Every village has its peasant-poet. They are poor poets, but good men, loved and dear to all. They produce a queer mixture of folk poetry and that other, artistic kind. Everything they write turns out to be impotent and grotesque. Having studied, quite poorly, other poets, such as Branko Radičević, Zmaj-Jovan Jovanović, or Aleksa Šantić, they tried to reproduce the thoughts and feelings of the people in this supposedly new style. They felt how feeble and unnatural they were, yet they had to write, and they were encouraged in this by the people, in whom still coursed the stream of folk poetry, drying up for lack of fresh wellsprings. Both people and poets seemed to suspect that something that had sustained them for centuries was irretrievably slipping out of their life. They feverishly sought to prevent it, but could not, and then tried to create something new. One could not live without poetry. This land may not be good for living, but it is fine for telling tales.

In the village next to ours lived two such poets, each different from the other, both good men, though secretly jealous of one another.

Poet Radoje was a lively ruddy little old man, loquacious and merry. He read his poems at village gatherings. He was a good *guslar*, too, but always had to be asked to recite or play the *gusle*—it was not proper for a poet or a *guslar* to perform without urging. When he had begun, he could be stopped only with difficulty. In his comic poems he would make sport of some happening that everyone knew about and that everyone would have forgotten were it not for his poem. He exaggerated, but never enough to make anyone really angry. His love songs spoke only of flowers, fairies, and moonlight. In one poem his nymph awaited him in the moonlight under a willow by the Plašnica River. They said this was a girl from the village on the other side of the river, whom now he remembered, and mourned for his youth.

The young people no longer enjoyed listening to *guslars*. The older men would retire to a separate room or by the fire to hear them, while the young people danced and sang their own songs. The old men will die, and there will be none to listen.

Radoje was also a better actor than the others when, at village gatherings, they would put on other clothes and make-up and act out various scenes from life in the village: spats and amorous scenes between an old man and a hag, or the troubles of an ugly and stupid bride. On occasion the performance was so convincing, despite all its crudity, that the spectators would forget it was all in fun and would, bit by bit, themselves take part in the repartee.

Sometimes it seemed to me that I, too, could write verses like Radoje. I would be walking, and verses would come into my head, and then, with a little effort, rhymes. But I never put anything down on paper. Who would dare? I would be discovered and ridiculed. Even my best friend, Mihailo, would make fun of me. He had inherited from his Communist uncle Milovan a tendency to ridicule poets and to look upon poetry as something worthless. Poets lie, he would say; their poems have nothing real in them. And on and on. Maybe Mihailo and his uncle were right. Still, poems are beautiful things, and one could hardly live without them.

It was not Radoje's poetry that was most interesting. He was even better at stories. All of him would become completely engaged in telling a story—hands, eyes, mustache. He was especially good at describing funny events, of which he knew a great many. After the war he told about my uncle Milosav. As a Montenegrin who boasted of being of the old stamp, Milosav refused to take cover in battle but fired from a standing position. While bullets sputtered all around him, he stood firm, as though they were flies. Some young fellow shouted at him from behind a shelter, "So they got you, Milosav, did they?" Milosav that moment fled for cover. Milosav was a bit angry at this story, but he never denied it. He would say, "It's not the same with Schwabs as with the Turks. You can't even get a good look at the Schwabs but have to fight hills and dales to get at them." Radoje knew how to depict the old-style Montenegrin bravado in the face of modern warfare.

Handsome men were marked—almost all had the reputation of heroes. Radoje had a story about such a handsome man. In Kolašin there lived the photographer Djukić, a man well along in years but still very good-looking, and therefore inevitably a ladies' man. He was wounded during the war, and his comrades carried him off the field. On the way they met a pretty girl, and the wounded hero called to her from his stretcher, "How are you, my little plum dumpling?" (Radoje pursed his lips and pronounced the word so that the first syllable sounded as juicy as a kiss.) On seeing this, the stretcher-bearers promptly dumped the hero off the stretcher, though his wound was hardly a light one.

There was still another anecdote from the war, a rather salty one. During the Battle of Mojkovac a woman brought her husband some hardtack. Because there was much snow, the wife put on some pantaloons. That night, the man and wife slept alone in a tent and began to make love. Since it was cold, and they were close to the front lines, the wife did not undress but her husband crawled into her billowing pantaloons with her. Just then some shooting was heard, and the husband lost his head as somebody nearby shouted, "Attack, attack!" All entangled and confused, the husband did not know how to extricate himself, whether from the front or from the back. His wife smacked him on the pate and cried, "The rear, damn your hide." This confused him all the more, and as he ran out of the tent, he shouted, "Attack to the rear, damn your hides!"

This joke is much too crude, but it pokes fun at the Montenegrin passion for attacking. It was one thing to tell a crude joke, and still another to ridicule old-fashioned heroism—and the older men did not like it. Nor did they like the profanity. Still, such jokes spread. So does profanity. Profanity, of the worst sort, came with the new regime, with the Serbians. All the gendarmes, all the petty clerks, chauffeurs, and butchers swore. And profanity spread. Everything comes and goes with time.

Even though these stories of Radoje's were not so nice, others truly were. These came from times that he hardly remembered. Time had purified and distilled them so that they were no longer rough or raw.

The best stories were about Lugonja, from the time

when the Morača had been wrested from the Turks, although Kolašin was still in their hands. Bands constantly crossed the Morača to the Kolašin side and back again.

At that time there lived by the Morača a certain dolt called Lugonja—big, clumsy, and slow in everything, as stupid as an ox. He was not a coward, but he always had bad luck; he was never successful as a hero. Poor as the poorest, a simple and happy fellow, he served as the butt of jokes, though he was well liked for his good nature and frankness. Unlike most guerrillas and heroes, he was punctilious and moderate in everything, and was remembered for this longer than many a hero.

Nenad Dožić, known as Little Nenad, was a famous guerrilla leader, a great and capable warrior, and a lucky one besides. One day Nenad and his band set out for Kolašin, and Lugonja insisted that he be taken into the band. All the men were against it, but Nenad agreed. He thought: Lugonja might be killed instead of some good man, and thus at least die with the fame of a hero.

The band spent the day in a forest, at a lookout spot and rallying place for descents on the Turkish settlement below. There they ate greasy mutton but had not a drop of water. None thought of looking for water until they had finished their job. Nenad assigned Lugonja to the door of a cabin, thinking that if the Turks spotted them, they would be sure to strike at Lugonja first. The dogs quieted down, and they opened the gate. Just then Lugonja noticed a small barrel of water. Forgetting where he was, the thirsty man rushed for the water. The barrel gurgled and gurgled. The Turk in the cabin awoke, thought an animal had upset the barrel, and rushed out. He spied the bandit and grabbed the luckless Lugonja by the arm. Lugonja, who was still in the throes of oblivion, thought Nenad was playing a joke on him, and he cried out, "Let me alone, Nenad, let me drink. My very soul is dried out!" Nenad, nearby, realized what the situation was and fired his gun into the air. The Turk released Lugonja and rushed into the cabin for his gun. But the alarm was given, and the bandits were forced back without even a yearling.

Lugonja had a herd of children. He never had a chance to get a bite to eat with them around. Once, his wife managed to obtain some flour and a bit of cheese. Lugonja noticed this and, in the morning, told her of a dream he supposedly had, and what it meant. "I must go," said he, "to Kolašin. This time I am determined not to come back without an ox, or a horse, or a good Turkish head."

His wife was more interested in his bringing back something alive than a Turkish head, but she was loath to put a hex on his heroism. She made him a poor mess of round cake. The children set up a howl. Lugonja comforted them. "Never fear, my lambkins, Daddy will come back." The children were not bewailing their father's departure, however, merely that he was leaving them without anything to eat. Lugonja paid no attention, but gathered up his weapons and provisions, and off he went—supposedly to join his outlaw band.

He did not go very far, just to the meadow by the Morača. It was a beautiful summer day, the shade beckoned him to rest, and so he fell asleep. Then he fished until night overtook him. He took a bit of round cake—and so it went, two, three days, until he had eaten it all. He came home empty-handed, realizing only then his predicament. What about the dream, the ox, the head, his wife asked?

"Ah, those Turks," Lugonja said, "those Kolašin dogs won't let a man even take a peek!"

Lugonja lives in those stories as a real person, a simple stupid man of strong passions, reared in a savage land at a time when only cleverness, heroism, and sacrifice were recognized as virtues. No clan would fight to claim him —he had not distinguished himself either through bravery or wisdom. Thus even his family name has been forgotten. He is remembered only by his nickname. And even that

will be forgotten. He represents the funny side of that
bloody and prolonged heroic struggle with the Turks. This
is not a land for jollity and merriment. They come only
briefly.

Besides Radoje, there was another poet in the village.
Milosav Šćepanović was a young man from a poor family.
He had left high school because of illness—tuberculosis of
the bones. He lay in bed nearly incapacitated, too poor to
go away for a cure. He had a beautiful face, dark wavy
hair, and black eyes made larger by illness. His lips were
unusually red, and his fingers thin and as tender as young
twigs, with the tips like buds, lively and supple. His face
was just barely shadowed with a mustache and beard, which
he rarely shaved, making him look even sicker and more
poetic. He was not too jealous of Radoje. He would laugh
and say, "Radoje is not a poet but a pleasant comic."

Milosav was the best student throughout his whole school
career. He knew French very well, even translating verses
from that language. His room was small and full of books.
He constantly read or wrote, in a very legible and beautiful
hand, to which he devoted much care, with initial capital
letters of fancy twists and loops. Perhaps it was his illness
that drove him to write poetry and to fight against sadness
and melancholy. He never spoke of death, but everyone
could see that he was looking it in the face. They said that
he knew he would die soon. And it was this everyone
found so hard to bear. They visited him often, especially
the young people. All the girls were sadly in love with
him. Nobody held it against them, for everyone anticipated
his death. There were no carefree moments for him, no
joy, and therefore the girls were free to love him. There
was no shame in loving the dying.

Why was he always writing and writing when he knew
that he would soon die? Why this chase after letters and
words? Certainly not just to pass the time away. Was it
because he wished to leave behind a trace of himself? Per-

haps only those who have come to realize that death is already upon them, and who have become so accustomed to expecting it that death has become a part of themselves, can possess such an inner beauty, such a gentleness of voice, and such a pitiful look in their eyes.

A certain professor took it upon himself to have Milosav's book of poems published. He waited for this, but calmly, without any great joy, as he did for death.

He remembered everything just before dying, not only every verse in his book and anything he had ever read, but every sound and color, every person he had ever met, every spring from which he had drunk, and every tree under which he had ever sat to rest. He died without seeing his book published. Everyone regretted this, but it was all the same to him. Having gazed so long at death, maybe he perceived something more important than his book.

There was a multitude of people at the funeral, especially young people. The day was humid, and because the villagers had come too early for the procession, the flowers wilted in the hands of the girls before they ever reached the graveyard.

Among the eulogists was Radoje, who, weeping, the tears coming from deep within him, read a poem dedicated to the dead poet. Death erases all evil between men, including Radoje's envy. Radoje was not able, and Milosav never had the time, to become what they wanted to be—poets.

The lowering of this coffin in the grave was the first such event to be engraved in my memory. The thud of the clods of earth on the coffin drummed all night on my own skin and on my memory. It drummed for me later at every funeral.

A third poet, an even more interesting man, used to visit our village. This was Arsenije Ćetković, Milosav's cousin, who had settled near Mojkovac, where he had a roadside inn. He was a special kind of poet, one who never either wrote or recited his verses, or regarded himself as a poet. He told tales about visions, so convincingly that it was difficult, especially for us children, to doubt their veracity.

He had a story about the Emperor Diocletian.

And God decided to liberate all lands and peoples, and to destroy the city. But He could not do this because of a righteous maiden from the Morača who served the diabolical Emperor and watched his cattle. She was the only sinless soul in the city, and, because of her, God pondered at length as He withheld His wrath from the city. However, His patience ran over, and He loosed lightning and thunder over the city, but commanded the angels to save the righteous soul. They managed to slip in and to transport to the Morača both the girl and a cow she happened to be milking at that moment, and even the bucket. The girl did not notice a thing until she found herself again on earth.

But the Emperor did not give up so easily. He arose from the ruins of the city undaunted and all the angrier and more terrible. He built up his city again, even bigger and solider. Unable to restrain the fiendish Emperor in any way, God sent His leading general and thunderer—Saint Elijah. The Saint and Diocletian wrestled on the land and in the clouds. The Saint was getting the worst of it, and so away he fled into the sea. Diocletian went in after him. The Saint jumped out, flew over and made the sign of the cross over the water with his staff. The water froze to a thickness of a hundred feet. Diocletian sprang from the bottom, broke through the ice with the crown of his head, and overtook Saint Elijah at the very gate of Heaven. He did not catch him, however, but merely caught a bit of flesh from the soles of his feet with his fingernails. That is why the soles of all men are so smooth.

Finally, the heavenly hosts subdued the Emperor Diocletian. But they could not destroy him, for he is immortal. They fettered him and cast him into the Morača, into the deepest hole, into eternal darkness. There they chained him fast. He tried constantly to tear himself loose. In order to make fast his bonds, blacksmiths came each Christmas to strike thrice on the anvil and to make new chains for Diocletian.

One day, however, the Emperor freed himself and emerged from the darkness once more to rule the world.

In his stories Arsenije mixed imagination with real details, the names of villages, mountains, and rivers the audience knew, and with everyday events, which made everything all the more convincing. It cannot be said for sure that he believed in those stories. It seemed that he did, though he added much that was his own. He loved to tell tales and he himself would thrill at them.

Most thrilling and convincing of all was one of his personal experiences.

One rainy and murky autumn night, Arsenije was in the mill on the Plašnica River near the bridge at Bakovići. The Plašnica was swollen and roaring. The torrent hissed over the flanges of the water wheel with the noise of a hundred serpents' lairs. The grinding of the millstone sounded grim and oppressive. The fire subsided to a glow, and the rain trickled through the roof. Arsenije did not know exactly what time of night it was, but it could well have been past midnight. When the grinding was done, he began to fill his sack with flour from the bin. Bent over as he was, he seemed to hear someone enter through the half-open door. He even heard a creak. He turned around. No one. A chill crawled over him, so that he almost lost his senses. Still he was not afraid. He continued to fill his sack. Again it seemed that someone had entered. Again no one. Again someone entered. Again only a yawning black-ness at the door as the rain and wind rushed in. Horror seized him all the more. He felt sick on recalling that others, too, had felt a presence in that same mill, and that evil spirits preferred pools, mills, caves, precipices, and dark nights. He had never put any faith in ghost stories. He had never before been afraid. There was nothing he could do. He said to himself that once he got to the field outside, he would soon get home.

When he was through, he took a flaming splinter to light his way and went out. The rain suddenly subsided, but a thick darkness enveloped everything in a black fog. He could not see his hand in front of his face. He did not mind, he knew the road well. He was born there and could have reached home with his eyes shut. He walked on and on. The terror he had felt in the mill was letting up. Suddenly, when he was in the middle of the field by the grave-yard, he noticed a huge black form, blacker than the darkness itself, coming toward him.

The figure addressed him by name, cordially as an old friend, and even asked solicitously, "Where are you going?"

"Home," said Arsenije. "Where else?"

"But this is not the way home," said the figure.

"How so?" asked Arsenije, confused. Suddenly everything seemed to spin around, and he could not recognize either the road or the large thornbush standing beside him. He could see that he had lost his way. He felt as if everything had become transformed.

The figure kindly offered to show him the way. Arsenije asked him who he was and whence he had come, but the figure evaded an answer, saying he was a wayfarer travel-ing over the earth, and that night had overtaken him. Arsenije joined him. Both kept silent. Arsenije did not know what to ask him. Not a question came to him out of the thick black night. The unknown stranger wore a cape, black and hairy, which reached to the ground, and on his head he wore a cowl of the same material. They walked and walked; Arsenije himself did not know how far and how long. Suddenly, Arsenije heard, as in a dream, the sound of rushing water, and started. How could there be any water when he knew that there was no water at all between the mill and his house?

The rushing sound and the place seemed familiar. He stumbled over a rock, and his torch fell forward. Just then he caught a glimpse of the stranger's feet. They were turned around, with the toes in back. Now everything was clear. The Devil was leading him on to drown him and to snatch his soul. Arsenije grabbed that rock and, with all his strength, struck the Devil in the back. The Devil turned around, stretched out his hands—his fingernails were like scythes—opened his mouth wide—his teeth were like the cogs of a millstone—and came at Arsenije.

"Hit me again," he said.

But Arsenije collected his wits. He knew that one could strike the Devil only once, and so he hurriedly crossed himself three times. The Devil helplessly ground his teeth and—disappeared. Arsenije then came to his senses completely and saw that the Devil had led him astray a whole hour's march, nearly to the next village and the old fort. Now he was really gripped by fear.

Arsenije went back. He recognized everything now. When he neared home he cried out to his wife. She answered his call, and he followed the sound of his wife's voice until he reached his house, more dead than alive. He fell into a fever, became ill, and barely survived.

When Arsenije came to the place in the story where the Devil opened his jaws and gnashed his enormous fangs, he would open his own mouth wide, stretch out his hands with his hairy fingers and long curling fingernails. Arsenije was himself all swarthy and hairy, with a huge head and long drooping mustaches through which one could see long yellow teeth. At that moment he would clamp shut his teeth with a loud snap.

It was far from the village to the high school, a whole hour's walk. One had to hurry to avoid being overtaken by the night at that very mill where the Devil had appeared to Arsenije. Small and hidden behind bushes, the little mill was spooky even in the daytime, full of unknown terrors.

Arsenije never got to tell all his stories. Early in the

autumn one year, he was killed by bandits in his own inn. People believed that it was done in league with the gendarmes. The bandits tied him up, and then fired into his broad hairy chest, and streams of blood spurted all over the walls. So said his wife, who was as black and as long as a shroud. Arsenije had no children. His house was deserted—except for a black banner on the roof and a black wife inside. All his stories died with him. No one else could tell them the way he could.

Good poetry lasts, like anything else man snatches from eternity by work and intelligence. Man, the most beautiful poem of all, passes away quickly and is forgotten. But man has not yet sung the last of that poem.

After I completed two years of high school, Father had me leave Aunt Draguna's and move into the town itself, where I stayed with a widow, Stana Jovanović. My elder brother, Aleksa, went to the normal school in Berane, and my younger brother, Milivoje, was still in elementary school. Thus in autumn of 1923, when I entered the third year, I had to go to school without my brothers. I had pangs of loneliness because of this, but also a spontaneous desire for independence. I had already passed twelve. I was no longer a child, but a young man.

The house in which I lived was rather large, pretty, and clean. In autumn and spring the swollen Svinjača tumbled just below. In winter the house sank in the cold softness of the snow-laden garden. Goodwife Stana, her daughter, Dobrica, and I lived in two rooms on the second floor, and a cousin of hers lived in the other two rooms with his wife and child. The ground floor served as a storeroom, with boxes of mulch and barrels of sauerkraut. This was the first house with two floors in which I had ever lived where the ground floor did not serve as a stable for cattle. This house lacked that unnerving and forced cleanliness that Draguna's son inflicted on her household. As in other houses, cleanliness here was not at all unnatural but came of itself.

Stana was a very thin and slight woman, like her sister, Stanija, who had married in my village, in Podbišće. They looked like one another, though Stana was younger, not yet forty. Her thinness and slight build did not bother her, or her sister, for that matter; on the contrary, she found them useful in going about her tasks all the more inconspicuously and efficiently—a real ant, completely. Otherwise, she was a woman without any distinction. A person would hardly notice her at all on the road, small and shy as she was. She was one of those people who live their lives unnoticed, and yet accomplish a good deal. Such people are quite rare, especially in a society that, except for intimate life, places everything on exhibition. Nothing about Stana was conspicuous, not even tenderness, which she showed to no one, not even to her own daughter and only child.

Whoever lived in that house for a while, however, would have observed that the whole being, the whole concern, and every movement of this woman, a war widow, was devoted to her daughter, Dobrica. A person would hardly suspect that so unnoticeable a love could be so unrelenting as to fill every pore and every thought. If anybody's life ever had a purpose, Stana's did—to make certain that Dobrica would grow up and live well and, if fortune would have it thus, to continue the family, and, in any event, to survive her mother. This was her only mission— a vow that this tiny, determined, and tireless woman made to herself and her husband, to mankind.

People are usually indifferent to those who love them too much. Dobrica was like this toward her mother. Besides, Dobrica was a city girl completely, while her peasant mother had married there near town. The two of them differed from one another in their habits and outlook. Stana, however, accommodated herself, again unnoticeably, to her daughter and her ways. Dobrica was aware of this love, but, being used to it, she accepted it as something natural and understandable. Had she not been truly a good child, modest and kindly, she could have easily wasted away the family property, if, indeed, anything could deplete Stana's calculated and fruitful efforts. Though Stana gave voice to complaints that a household and property go to ruin without a man's hand, this was only an expression of love and respect for her fallen husband. In fact, she cared for the property, the house, and her daughter better than if her husband had been alive.

In this, too, Stana was like her older sister, Stanija in Podbišće, a war widow with two children, who not only reared her children, but enlarged the property and built a new house. Stanija's son, my schoolmate in elementary school, became a Communist. Even were it not for this, he would have stood out among the villagers for his intelligence and sobriety. Stanija helped not only her son, but other Communists as well. What else could she do, when her only son was a Communist? She spread illegal literature, joined demonstrations, old as she was, organized protests against arrests, fed and hid fugitives. She continued to do this during the war. Her son lost his life, the house was burned down, the cattle were plundered, and the garden destroyed. She herself was wounded. Only an evil hand could have fired a bullet into this little creature, who was all skin and bone and a little bit of flesh.

After the war, Stanija, who could never grow so old as to lose what was most vital about her, a ceaseless building and rebuilding of life, was indomitable and fearless. She rebuilt everything—her house, garden, and property. However, she would not join the village collective farm, despite her love for the Communists, which she had sealed in blood. "I can't, I won't; they don't work as they should there," she argued. As some thirty and more years ago she had reared her children, so now she began to devote herself to her grandchildren. She managed to send two grandsons through school. Life gained new meaning, finally, after two wars.

There are such Amazons, whose strength and intelligence find expression only when they are left to themselves in the struggle with a bitter life. Such were these two sisters, Stana and Stanija.

Being less forceful, Stana was not as lucky as her sister. All her efforts went to nought. She, too, sent her daughter, Dobrica, through school. She found her a good match. But Dobrica and her two children were blown to bits by an aerial bomb at the very beginning of the last war. Her mother did not even have a grave over which to weep.

Passing frequently through Kolašin after the war, I often had a wish to visit Stana. She was still alive, but none of my friends or acquaintances, and these were mostly local officials, knew anything much about her. Evidently she lived a retired life, lost in loneliness and in memories of Dobrica, left to a life that had been demolished to its very depths.

She never came to see me, though she must have known of my arrivals. On the other hand, I did not seek her out because in the circles in which I moved no one would have understood my visit, being what I then was, to a woman who had no relations with the Communist powers that be, and who had never found it possible to be even an active Communist supporter. Rebelling inside, I nevertheless conformed to the prejudices of the closed circles of the Communists, who see only themselves and their charmed world and frozen ideas.

Whenever I happened to go by Stana's house, I felt the reign of death there. Yet even then everything was clean and orderly. The lonely little woman was most probably wasting away in a pure and quiet sorrow. Probably only things still bound her to life—fruit trees and vegetable garden, and the pale yellow flowers on her window sill. Everything was permeated with an ineradicable tragedy. There was not a trace of a live and vital human being. The windows looked black, and the courtyard showed no trace of work or life. War had not destroyed this house or garden; they had been ruined by a life of shattered thoughts and hopes and wishes. Everything had begun to wear out, to crumble and to rot, either to waste away entirely or to await another master who would infuse the strength and freshness of a new life and joy into this house and everything around it. Yet it was precisely here, in this house and garden, and by that wall behind the garden, that my first boyish loves and sorrows, my first verses and first disappointments and rages against human misfortunes and inhumanity and misery were conceived.

The memory revives dear familiar objects, and awakens forgotten thoughts and feelings. Yet all slowly falls to ruin together, sinking into oblivion and nought.

There were in the vicinity thousands of excellent places for a city, each more beautiful than the other. Kolašin was erected on the most beautiful of all, though it was not a desire for beauty, but human misfortune, both Turkish and Montenegrin, that built it.

It seems that there never had been a real fortress there. The Turks began to concentrate at that spot as early as the seventeenth century and to build their towers, and in 1648 the settlement was surrounded by a wall. In contrast with the majority of Turkish settlements, the site itself was not on a river, the Tara in this case, but on a rise near the sources of two rivers, the Tara and the Svinjača.

It was most convenient to erect the walls there, it appears, because it was so steep on two sides. Thus the settlement began on an elevated and commanding height, on a plateau surrounded by hills, supplied by water, and washed by swift streams. It was impossible to count the mountains and hills around. Valleys and streams converged on Kolašin from all sides. Fresh mountain breezes came from everywhere and butted into one another over the town, as though the battlefield below were not enough.

Kolašin was settled by powerful and militant Moslem clans. It was constantly attacked by the surrounding Montenegrin clans, who boasted of sacking it many times. This is why Turkish commanders would ride out of the gate to pacify the surrounding rayah and to bring back the heads of disobedient rebels to decorate their towers and walls. Kolašin was the prey of robber bands both day and night, for centuries. The rayah called it bloody Kolašin, while Turks all the way to Istanbul praised its heroism, its water and air, and its fair maidens, with their red lips, black hair, and shining eyes.

Many roads met in Kolašin; it was the knot that kept apart the various Montenegrin clans and lands, until Commander Miljan Vukov razed it in 1858, putting to the knife all the males he could find, except for children, and expelling the Moslem population. Later some returned to their homes, but it was never again to be a Moslem settlement. The Moslems continued the struggle with the Montenegrins a few more years, until 1878, when, after the Congress of Berlin, the whole region fell to Montenegro.

Then the Montenegrins—the Morača and Rovči clans— began to divide the Moslem lands and to found a new city on the same site. The Moslem houses and mosques had already been demolished, and their graveyards leveled to the ground, as though they had never lived and ruled there. Neither in the town nor in the entire region was there a single Moslem left, except for some gypsy blacksmiths, useful to the Montenegrins, who held it beneath their dignity to engage in such an occupation.

The town managed to rise rapidly and in good order, with a market place at the center and streets that converged from all sides. The houses looked like those in the village, maybe a little better, built of rock and roofed with tile. The people who settled it were erstwhile peasants just beginning to engage in the handicrafts and in trade. The war lords, especially the older ones, were reluctant to live in town. They expropriated the best Moslem lands, erected towers, and preferred to live by themselves, in their gardens, and from there they climbed the hills, herded their cattle, rode their excellent horses, and passed judgment on their people.

The little town had barely two thousand inhabitants, but it was the heart and soul of the whole region, a spider that had spread its web to take in the most distant settlements, to suck out their strength. By itself the town was powerless. Its men were not distinguished for anything in particular. It was a center for everything—government, trade, and culture. This is what gave it strength. People had to gather somewhere, so they gathered there, and from there they spread abroad news and wares and new ways.

The peasants hated this little town, more perhaps than in Turkish times. Under the Turks it had stood for might and lordliness. Now there was neither. Still, the peasants themselves could no longer manage without the town. Its strength lay not in weapons, though partly so, but in the fact that through it passed those vital arteries that nourished the villages while at the same time sucking dry their strength.

Life in the town was easier and better. Yet even there one could see how much the peasant had to scratch and scrounge to earn a penny. Outside the town the peasant was spare, ragged, the owner of a miserable pittance, which he had to sell, resentful of the market place because it skinned him, yet ready to grab whatever loot he himself could get. How could a family live that had five or six members, one goat or scrawny cow, a sliver of unfertile land, and nearly no other means of support? Yet somehow they lived and managed. True, they ate heartily only on the feast of their patron saint or on Christmas; they ate meat hardly at all, they mixed flour with cabbage; but they survived.

Misery in the villages produced mobs of workers for the construction of the new road from Kolašin to Mojkovac. They came from all sides. Many could not get work. The others dug the ground and pounded rocks into gravel, selling their labor to the employers for extremely low wages. The road progressed slowly, a few kilometers every year.

There was no money. The peasants used to say, "This road will get several parliamentary deputies elected."

The employers were an arrogant and merciless bunch. They were for the most part short pudgy men, clever, coarse, and unyielding. They knew the workers inside out, and accepted only those on whom they could keep a tight grip. They paid by the piecework system, inspected the work done, and demanded speed. Somebody else in turn, even stricter, supervised them. It is a marvelous thing to see human hands fashion a road out of mountains and cliffs. All white, that rocky rope slowly unwound along the wooded banks of the Tara. Ragged, lean, and hungry men, dusty with rock powder, and those rocks were locked in a shattering contest in which the men strove to survive and to maintain their families.

Why must men do that? They toil, they suffer, they beget children, they die—and others after them do likewise —as though this was the only meaning in life—to work and to die.

Everybody expected some benefit from this road—both those who worked and those who supervised, not to speak of those who promised their constituents hills and dales of gold.

It was rumored that a new mine would open at Brskovo, where a mine had existed in medieval times under the Nemanja kings. The forest would be cut down, it was said, and everything would come to life. A fever gripped men at the thought that a narrow white road would penetrate the ravines. Actually, very little changed in those places the road reached. But the teamsters disappeared, with their drawn-out songs and the sad tinkle of the little bells that hung from the scrawny necks of their horses.

It was not easy even for us, with only Father's pension. What it must have been like then for the peasants! There were years of drought, as always in time of need. Frequently there was no grain to buy. And when it came, the

price was high. Even in our household there was great joy
when Father brought in a load of grain; we were safe till
spring and would not lack for bread.

In the villages there were usually only two or three
families that were better off; the rest were all paupers, or
else sharecroppers. One could immediately spot a well-to-
do peasant. He had a riding horse and dressed in better
clothes—worsted trousers and a gold chain. Such men had
a calm, radiant look, and when they shook hands they did
so as though they were distributing alms, as though to say,
"See, we could be otherwise, but we are not big-headed."
Their wives and daughters were prettier, too—fuller and
fresher, and stronger. Differences among the peasants were
even more obvious when they gathered in a crowd on
market days in town. The more well to do dressed in their
very best on that day, in all their finery. The poor were
never able to conceal their poverty, which was all the more
striking in comparison with the well to do.

Market day in Kolašin brought together all the miseries
and woes of the peasantry. In torn homespun coats, without a shirt underneath, with gaping shoulders and elbows,
the peasants wandered about without anything to do or
listened as the more well to do discussed politics, in hopes
of hearing something that might be useful. Thin, angular
women, with protruding joints, crouched over a pot of
curd, a basket of eggs, or a jug of milk, waiting from
morning for buyers who would not come until the afternoon, when the price would fall half a dinar.

The cattle at the market looked the same—scrawny and
worn out, with alarmed looks and listless gait. Those who
grew a bit fatter in the spring rarely got to market; the
buyers always resold them somewhere farther away. The
butchers boasted that their meat was prime, just as the inn-
keepers boasted that their plum brandy came from the
Morača. It was rarely so in either case.

The town was still a young one, without even a row of

trees to cool it. All the townsfolk had relatives in the villages, and if the town were suddenly to be destroyed, everyone would have found a place to live with village kinfolk. All the inhabitants in the town, except the Marić family, had been peasants only yesterday. It seemed, however, that the townspeople hated the peasants all the more for this, and had a special contempt for them. This was hatred through contempt. As soon as anyone moved to the city, he considered it his prime and most sacred duty to hate the village. Something seemed to snap inside of him, as it might in changing religions. Most of the townspeople still kept animals and tilled small plots around the town. They were contemptuous of the peasantry. Yet, when they went into the villages, they were indistinguishable from the peasantry except for the fact that they wore city clothes more and were—though not all of them, at that—somewhat cleaner. They were petty shopkeepers, coffeehouse owners, and artisans. The blacksmith gypsies were outside of town, by Pažnja Creek. One could always hear the hammers from Pažnja. They mended plowshares, harrows, and axes—and the chains of the dread Emperor Diocletian.

If the townspeople, erstwhile peasants, had contempt for the peasants, the peasants, in turn, hated them, the way animals hate traps, traps whose location and purpose they know and yet cannot escape. The peasants looked upon the townspeople as a sluggish, wily, and lying breed, who eat little and delicately—fancy soups, tripe, and pastries— and waste away in damp, crowded little rooms.

The town and village children hated one another, too, with an unconcealed hatred. The town children used to beat us peasant children and, in the afternoon as we set out for our homes, they would throw stones or icy snowballs at us, depending on the season. They stuck together more than we did, and attacked in more united fashion; in this they were more skillful. We peasants distinguished ourselves with solitary feats and for our slow brute strength.

There was a rich family in town, the Marići. They did
not hate anybody. For them all people—both peasants and
townspeople—were just things that could be manipulated,
profitably or at a loss. It seemed as though no one hated
them either. They, too, seemed to be regarded by others as
things. But they were resourceful and active people, who
could agree only in manipulating others.

Besides a big store, the Marići had a beautiful house that
looked over the town and had a large garden behind a wall.
They also had a sawmill on the Tara. They no longer lent
money at interest, and gave goods on credit only to reliable
customers.

This family lived a secluded life, behind fences and walls,
within a garden whose dark coolness beckoned in the summer and whose fruit trees spread their fragrance in the fall.
They cultivated only a few friends. No one knew their
wives and daughters.

Everybody believed that whatever the Marići slept on or
ate must be something special. The whole region stared at
their mansion—but saw nothing. "Not even if you were as
rich as a Marić . . ." was a local byword. As merchants
are, they were nice to everyone. Like merchants, too, they
kept their secrets.

They were not very interested in higher learning. In
their physical make-up and movements, and in everything
else, they were soft, languid. Their women were like that,
too, pale and plump, and serene at all times. Marić festivities were never noisy and were apparently held merely to
observe custom. So, too, when one of them died, they
buried him without wailing and keening, with only stifled
sobs and tears. Leading easy lives, they made terms with
death, also, more easily.

It was said that they were not even Montenegrins—because they were such good merchants—but of Illyrian
stock, of that people who inhabited these lands before the
Slavs ever came, and who left behind only graves and the

names of rivers and mountains. The Emperor Diocletian had been of that stock. His empire and the Illyrians are now a part of the distant past. So will the Marići be someday.

The Marić properties were divided when I was there, but, just a short time before, they had been the joint property of two brothers, Akan and Antonije. Antonije left many sons, whereas Akan had an only child, Tošo. Tošo made a settlement with his cousins. The reputation of the house did not wane, however, nor did their commercial ties with all of Montenegro weaken.

Akan had been a man of parts. He had established and enriched his family. Whole legends were spun about him, and even his sayings were still repeated, especially those dealing with parsimony.

According to legend, Akan fell into wealth in an unusual manner. The girl Akan married was the last and only heiress of an old non-Slavic family. From generation to generation a document had come down in her family describing the location of a buried treasure. Since the family line ended with her, she turned the document and inheritance over to her husband. Afraid that if anyone else had to be called to read the document, that person could send him to the wrong place and grab the treasure for himself, Akan traveled to Turkey, to a town that knew nothing about him or his country, and found a man who could read it. Then he hurried back, traveling day and night. One night he went with his brother to Svatovo Cemetery on Mount Bjelašica and began to dig at the appointed mound.

Now on a bare and quite noticeable mound stands a stone tomb, in which the treasure is said to have been buried. Wrapped in the mystery of ages past and of legend, this place became known, even during Akan's lifetime, as Akan's Grave, as though he who had discovered the treasure had buried himself there.

Nobody ever denied the truth of this story. Not even the Marići denied it, though they did not confirm it either. They kept still, as befitted true merchants when their property is involved. Whether the story was true or not, Akan became rich by his own efforts anyway. He got to Kolašin as soon as it fell to Montenegro in 1878, when it began to be a real city, with shops and inns. Akan was one of those early merchants who lent money at interest, traded in cattle and meat, and slowly introduced the sale of manufactured articles. Such men—the first Montenegrin merchants—appeared in all the towns of Montenegro after 1878. Until then the tiny Montenegrin state had lacked towns, and, thus, merchants and artisans as well.

Akan's business was a demanding and even dangerous one. He had to travel to Turkey along unsafe and bad roads, and haggle with teamsters and bandits as well as with the corrupt and arbitrary Turkish authorities. He also had to contend with the wild and rough Montenegrins. He put up with it all, however, and founded a family and fortune.

Despite all this vast wealth, Akan died, they say, like any other man, eager for life and its joys. Just before he died he said, "In the sight of death all wealth is nothing." Perhaps he meant to say that in death all men are equal. Nevertheless, Akan died a man who was respected for his diligence and ability to save money, both uncommon in these parts, and because he always kept his word, very common here. Nobody in the family carried on his business, at least not in the same way or with the same skill. They became merchants like all the rest, only richer.

Akan's son, Tošo, lived the life of an eccentric. He completed the school of commerce, but never did anything, living from the income derived from his inherited wealth. Yet he was not one of the spendthrift gilded youth. Rather modest in everything, he lived amply and comfortably, without luxury and show. It was as though he had calculated his whole life, not only in years, but to the very day, and apportioned his income to last him to his final hour, down to the last penny. He had no passions or great joys. He never married. He simply lived aimlessly, as though he wanted to live out his life as calmly and comfortably as possible.

The other Marići, though they sought to maintain good relations with the Montenegrin authorities, still imperceptibly looked upon them as strangers, as a phenomenon which comes and goes, while the merchants and their business were here to stay. Tošo, on the other hand, became completely the Montenegrin, in dress and speech and habits. He looked like one, too—large and big-headed, only a bit fatter and slower than ordinary Montenegrins because of his easy and carefree life.

It looked at one time as if he, too, would be seized with the fever of opening mines, cutting down forests, and starting new ventures. My father, with his boundless imagination, encouraged him in this.

As were all other Montenegrin authorities, Tošo was on good and friendly terms with my father. Father talked him into looking for mineral deposits on our property, of all places. Father talked and talked about the medieval Saxon colonists who were brought to mine there. He tried to guess where they got their water, where the mine shaft might have been, and where the veins of ore might have led. Tošo, who liked to eat *prosciutto* and cheese and drink mild brandy, would sit lazily, hardly uttering a word, listening to Father late into the night, obviously as disinterested in mineral wealth as he was in everything else in life. They hired laborers, who worked some five or six days. Then Tošo dropped the project and returned to his previous motionless existence, whose ponderous delight he alone was capable of understanding. It was as natural for Tošo to abandon that venture as it was for my father to get over his mining fever and to latch on to some other idea.

Everyone keeps coming back to what is the essence of his life.

Not only the townspeople and the authorities, but the peasants as well were occupied with political discussions. The Marići, however, were somewhat different; they remained apart in this. They always kept on the good side of the authorities, making no distinction between politics and business, conscious of the fact that in this country whoever opposed the authorities could only lose. They were afraid to mix their wealth and connections with politics and to use them to influence the authorities or the political loyalties of other men. All political struggles simply bypassed them. Moreover, had it been up to them, they would have preferred to have none of all the troubles and uncertainties that war and political disorders bring in their wake. From their point of view, it would have been best if the world were divided only into those who buy and those who sell, those who work and those who are idle, with no authorities or politics except what was needed to put down looters and robbers.

There was still another prominent merchant family in the town—the Boškovići. Though they were Montenegrins, their life, too, was secluded and isolated, though not as much as with the Marići. The girls in their family were famous for their beauty, as healthy as peasant girls, and yet as soft and gentle as city girls. One of them, Julka, went to high school with me. She was bigger, almost a young lady, though only fourteen. She was all aglow, with big black eyes. Her breasts were already full, and dimples were appearing on her arms. She was not very good at learning, but no matter; she would marry well, with a dowry. She kept away from us peasant children. Whenever she did speak to us, she was cordial and charming; her smile showed her large regular and unusually white teeth. She was the only child in school who washed her teeth at all.

Glistening amid the mountain heights and streams, the

town looked sad at night, with only a few lanterns and many taverns in which jobless men—and there were many such—killed their boredom, mourned their past, and dreamed of the kind of beautiful and carefree life that men have in some distant lands.

Are there such lands? And is there such a life?

The townsfolk lived for themselves, and the peasants for themselves, even though both dreamed of a better life and were bound to one another through trade and politics.

The Montenegrin chieftains, it seems, no longer dreamed dreams, and were lost both in life and in politics. Nearly all of them lived in villages, yet they liked the comforts of town life. They were not merely in between the burghers and the peasants, but in a class by themselves. Probably the differences that separated all of them—the peasants, the burghers, and the chieftains—were less evident in their way of life than in their psychic make-up. They differed more in what they wanted than in what they were. They stood poised against one another, and this deepened the differences even when there was no reason for it.

The town would have been colorless if on Mondays, which were the market days, only the ragged peasants came. With them, however, came the pensioned chieftains from the villages, each on his war steed and in ceremonial gold and velvet costume. In a moment, as they shone forth in the market square, they restored to the town some of its lost, bygone brilliance.

Miloš Dragišin Medenica could easily have been the most renowned chieftain in the whole land had he not been so retiring. He was of medium height, of sturdy and well-knit frame, and had mustaches so yellow that they blended into the gold of his breastplate. He was in all things deliberate, slow, gentle, and not proud. He even talked with the children. Yet this very man, who was distinguished from the peasants only by his somewhat richer dress, had been the main protagonist on the Montenegrin side in the Battle of Mojkovac, between Montenegro and Austria.

The entire front was commanded by Division General Janko Vukotić, a man of indubitable military talent and such good fortune in battle that his soldiers were convinced that wherever he was there could be no defeat. Miloš Medenica was not the only commander in the Battle of Mojkovac. It was, nevertheless, his Kolašin brigade that took the brunt of the battle, so that the command was principally in his hands. During the days of that battle, without which the history of this land would have been spotted with shame, the indomitable resistance of a handful of mountaineers against a mighty empire found expression in Medenica, an unassuming and modest man, in fact a peasant. He was like all the rest. And, like all the rest, he had merely carried out a task assigned to him by his superiors and inherited from his forebears. He was not a man of any great learning. He had gone through the military academy in Cetinje, and the rest he learned in battle. Montenegro was invaded by a modern army, led by officers of the Austrian General Staff. Against them was he, Miloš Medenica, almost a peasant, leading starving, half-naked, and poorly armed and ill-equipped peasants. But the Austrians were invading a foreign land, whereas these peasants were defending their own.

The Battle of Mojkovac was inscribed in fire and blood in the first memories of my childhood, and became quickly transformed into a terrible tale of an inescapable and savage, yet purposeless, slaughter for the homeland. The cannon awakened each morning, and at evening gave no rest. The machine guns barked all night and bit into every word and every dream, leaving untouched not a single bud on the branch. Women wailed day and night as the dead and wounded were brought in. But the Austrians fell like flies, and whatever they won during the day they had to give up at night. The wolves fed on their flesh all winter in the ravines, and left the villages alone. For years later, horrified shepherds would come across bones in ditches and gullies.

The battle was fought during the first three days of Christmas, January 7–9 by the Orthodox calendar.* The entire nation celebrated these days in blood and death, without even kindling the Yule log. In spite, the Austrian command would not let the Orthodox celebrate their greatest holiday in peace. They did not suspect how much fiercer the resistance would be because of this.

Much has been said about how through this battle the Montenegrins saved the Serbian army, which was retreating to the sea, from being cut off and captured by the Austrians. This has been meant as a reproach to the Serbians, who were not only ungrateful to Montenegrins, but called us traitors. To be sure, it was not quite so, at least not entirely so. The main Serbian army, with the cabinet and the court, had already got away and were near Durazzo and Scutari at the time of the Battle of Mojkovac. But before that, the retreating Montenegrins had been able to relieve the pressure on the Serbian army from Bosnia.

The Battle of Mojkovac was a purely Montenegrin fight —the last and the most glorious in the history of this small state. With it, the state was extinguished, in a final bloody and unforgettable flash of heroism, glory, and legend. The Montenegrins retreated and retreated, from the Battle of Glasinac in Bosnia to the threshold of their homeland proper, Montenegro, whence they had started in 1912 and in 1914 to wage war on ancient and great empires. There could be no further retreat for that army. Further was Montenegro, honor and glory, the past and a life for which

---

* Some Orthodox peoples still use the Old Style, or Julian, calendar, which places Christmas thirteen days after its date in the New Style, or Gregorian, calendar.

one's blood must be shed to the last drop. Naked, hungry, they strove to survive, though all, down to the last soldier, knew that they could not prevail. They did prevail for a moment, at least in battle, in slaughter, where the real strength of a people and an idea could be measured.

Behind the back of this martyred and bleeding army, through the cowardice and speculation of the court circles, there fell, almost without a struggle, those symbols of independence and liberty—Lovćen and Cetinje.* The invincible army of Mojkovac found itself without any backing, and felt betrayed. Montenegrins fought without surcease at Mojkovac and fell in hecatombs, while at Cetinje their state was collapsing because of the treachery and intrigues of their leaders. At Mojkovac, the history of this land was being enacted in a bloody last stand, while over there everything was falling apart as though there had never been a past or the desire to survive for the sake of some distant generation even now dying on its own threshold and hearthstone.

There was another reason, however, for the shameful, unmilitary fall of the Montenegrin army. This was the sinister and moot role of the Serbian government. This government, and its representative in the Montenegrin high command, Colonel Petar Pešić, not only did nothing to overcome the pusillanimity of the Montenegrin leaders, whose whole policy, in both war and peace, had already boiled down to the bare preservation of power, but they encouraged it. It was the Serbian government that saw to it that the Montenegrin army did not retreat with the Serbian, to make sure that at the end of the war there would not be two armies and two dynasties, which would have complicated, and perhaps made impossible, the establish-

* See note on page 49. Cetinje was the capital of Old Montenegro, the center of its resistance against the Ottoman Turks. Lovćen is the mountain that overlooks Cetinje from the north, and it is celebrated in Montenegrin folk poetry and in Njegoš's *Mountain Wreath* as the bastion of Montenegrin freedom.

ment of a united state. Resentment against the Serbians, brethren, remained long after, bitter and deep. Was it necessary to put to shame our glory and honor? Was it necessary for those who had obtained their weapons for over three hundred years by taking them away from the enemy to surrender them now? A sickly shadow fell over the wonderful dream of unification before it had ever been realized.

Of course, the Serbians were not to blame that our own Montenegrin leadership was rotten and wayward. Yet they could not have pushed us along the path of a shameful defeat, which wounded the soul of a whole people, without placing on themselves the mark of Cain.

Must all great things be achieved in a dirty and vile manner? Apparently such is fate—at least the fate of this country.

The grandeur of the holocaust at Mojkovac was not in victory, for there was none. The enemy was simply stopped, while the state dissolved at the same time. The grandeur of this battle lay in the expression of an undying and inexplicable heroism and sacrifice, which held that it was easier to die than to submit to shame—for in death there is neither defeat nor shame.

Greatness demands greatness in everything to the very end.

Miloš's adversary in the Battle of Mojkovac was the audacious and extremely courageous Colonel Rendel, a real Austrian. A proud noble, officer of an omnipotent empire, he simply could not imagine that he could be stopped by an opponent who was weaker in every respect.

In heavy battle Rendel occupied the heights overlooking Mojkovac and thus opened up the way along the Tara Valley to Kolašin. But he had to keep those heights. They were the only passage, for on one side was Mount Bjelašica, impassable in winter, and on the other side was the impenetrable chasm of the Tara and the even more forbidding Sinjajevina. Here was the gateway to Montenegro.

Vukotić came to the front and ordered that the enemy be repulsed. The Montenegrins began their desperate charges. Three battalions of recruits, which had already been decimated in earlier battles, were now decimated for the last time. They had gone to war with a thousand soldiers each. They dwindled in number from encounter to encounter, so that after the Battle of Mojkovac they had less than three hundred soldiers each. Other units hardly fared better. But the Austrians were repulsed. Then, in a rage, Rendel bared his sword and led his own unit in a charge. That won him the coveted Order of Maria Theresa, but not the battle. Both the Austrians and the Montenegrins remained on the same heights overlooking Mojkovac, in forests and in snow, facing one another at arm's length and massacring one another day and night.

So it continued until our army retreated without a fight after the unfortunate fall of Lovćen. With the Battle of Mojkovac the Montenegrin state fell while Montenegrin arms flashed in their final brilliance. That battle was the pride of those men who fought it and of the nation to which they belonged, and even of the Montenegrin chieftains.

After the war, the chieftains found themselves in a situation as though they had never fought. They received pensions, to a man, but they fell out of joint with the times that followed, and were ignored and discarded. Neither the soldiers nor the Montenegrin army as a whole received any recognition. The Battle of Mojkovac was not given recognition either, hardly recorded, like the blood and martyrdom of the Montenegrin army. Yet there arose and remained a legend about it.

People respected Miloš Medenica as a brave and stalwart man, but paid scarcely any attention to his role at Mojkovac. He held himself modestly, as though unaware of the greatness of the task that history had assigned him and that he had carried out faithfully. He lived in retirement on his property, forgotten, as the other chieftains were. And like most of them, he, too, adapted himself to the new order of things, though he could never make peace with it.

Miloš's father, Dragiša Perkov, was still alive. He, too, was a renowned hero and chieftain, one of the leaders in the War of 1875 to 1877. An old man of slight frame, dressed in the tunic of bygone days, he walked with difficulty; the children laughed at his old age, at his trembling mouth and hands, at his deafness and poor eyesight. When Dragiša died, his son Miloš dressed him in ceremonial clothing—the Montenegrin costume. Late that night, however, robbers opened the grave and stripped the old hero of his costly raiment. Miloš's family hid this as a shameful secret. The whole region was ashamed. That is what happened to the man who led the Battle of Mojkovac.

Not even the dead are spared when the living become lost and grow evil.

Dragiša Perkov was one of those chieftains who received Turkish land after the liberation of Kolašin. From their poor homesteads the chieftains had moved to their new lands, whose quantity and quality were determined by the rank and fame of each chieftain. They worked their new estates with the help of hired hands. Though hunger for land was great and general, there was little land to be had— so little land that there was no place for tenants. Veteran soldiers of the campaigns of 1876 to 1878 had received smaller and poorer parcels. But they were free on their own land.

The chieftain of the Kuči, Marko Miljanov, on the other hand, had not rushed to get any property. Even more than other Montenegrins he was proud, openhearted, and unselfish. He was the greatest hero of his times. It was claimed that he had cut down with his own hand over eighty Turks. In times of peace, however, he was distinguished for his humanity and unselfishness. He had quarreled with Prince Nikola, and retired to a small former Turkish farm at Medun, which his soldiers had given him as a gift. No longer an officer, he spent some ten years there amid the solitude of the rocky crags, tending his vineyards and bees, learning to write so he could record his reminiscences of heroes and thoughts on the transitory nature of everything but a man's integrity, which must be preserved and which must prevail at all costs.

The expropriation of Moslem lands after the War of

1875–1877 had taken place, also, because the Moslems themselves left their lands. They, the ruling caste of yesterday, would have found it intolerable to live under the Montenegrins, even if the latter had not driven them away. After the War of 1912, however, the Moslems all remained on their lands. The Montenegrins then gained their lands through purchase. True, the price of the land fell because of looting and violence, but there was no longer forcible expropriation.

The officers in the War of 1875–1877 died out gradually, naturally, while the officers in the War of 1912, still in the prime of life, were cast off by life, that is, by the new regime, to make way for younger but colorless and unpersonable servants of the new, Yugoslav, regime.* The former were superannuated; the latter were superseded. Among the former were men whom age had made humorous; the latter were only unhappy.

The younger officers were very prone to anger and were quick to reach for their weapons. They did so, however, only in quarrels of a personal nature, matters involving wounded honor or newly gained property. They had lost the ability and strength to offer any resistance to the new state of affairs. They were capable only of maligning the new powers and making difficulties for them such as they themselves had hardly ever encountered. Quarrels, usually bloody ones, broke out over practically nothing, most frequently over a word. Words became very important indeed to these men, all the more so, it seems, be-

* The "Kingdom of the Serbs, Croats, and Slovenes" was first proclaimed on the island of Corfu in July 1917, a union whose sovereign was King Peter I of Serbia. From 1918, when the unified kingdom became a reality, until his father's death in 1921, Alexander acted as regent. From 1921 to 1929 Alexander was "King of the Serbs, Croats, and Slovenes," then in 1929 he proclaimed a dictatorship and changed the name of the kingdom to Yugoslavia (in Serbo-Croat, "land of southern Slavs"). King Alexander was assassinated by Macedonian and Croatian nationalists in 1934 while visiting France. His son Peter II succeeded, under the regency of Alexander's cousin Prince Paul.

cause their lives were deprived of any substance. They
lived for the sake of talk and died for it.

In Bijelo Polje there took place an inconceivable settling
of accounts among these former officers. Having fallen out
with each other over sharp words, wounded pride, desire
for power, and soured against one another largely through
the bitterness of their lives, they split into two groups, as
though by arrangement, and one night came across each
other in the tavern of a certain widow, in a tight little room
on the second floor. They were separated only by a very
long table, at which there was not room enough for everyone.

On one side was Simo Terić, a slight and slim man, but
an aggressive fellow, who was very brave when it came to
words. He was at the time a district chief in Šahovići. The
officers in Bijelo Polje, who had challenged him to meet
them, had thought he would disgrace himself by fearing to
come to sit with them and have it out. If he were foolish
enough to come, they hoped to kill him in the ensuing
melee. But Simo was not frightened off. Indeed, he prepared himself well for an encounter. He wore two re-
volvers. He came to the meeting place with a bodyguard
and two friends, determined to lose his head rather than be
branded a coward.

Then began the taunting. All their passions burst out in
words. The words became too hot. The first bullet—they
say it was fired by Simo's bodyguard—extinguished the
lamp. A furious firing raged in the darkness. Over two
hundred bullets were fired in the course of several minutes.
The wonder was not that there were so many wounded,
but that anyone at all came out of that inferno alive.
It seems that the proud Terić, ready for anything and
knowing what they were cooking up for him, was the first
to begin the quarrel. He dropped to the floor in time to
avoid the hail of bullets which struck from all sides. When
the firing died down and the considerably diminished

enemy force had fled, he jumped out the window, found a horse, and got away to his own district. Bad blood and court proceedings dragged out over this affair for years afterward.

There were many lost causes everywhere around, especially in the villages. If they were more noticeable with respect to the officers it was because they were men of reputation and, moreover, the wisest men in that environment. They presented a picture—somewhat different with each individual—of men belonging to one age who were irrevocably disappearing in another, different age.

Some of these officers drank, others gambled, and yet others chased after women—for everyone goes to ruin in one of these ways. Yet there were exceptions to the rule. Some became obsessed with the necessity of getting rich. Others fell prey to the passion of hunting. Still others preferred to sit and engage in endless conversations.

There also came into being a certain class of women who could be easily recognized by their mincing walk, grimacing smiles, and confident air in male company. They, too, were lost in their time; their one thought was to live out what remained of their unfulfilled youth and life as fully as possible. One could not say that they were dissolute women. No one even regarded them as such, though they were free in their behavior and threw themselves into love affairs without restraint. Their amorous experiences were rare, always hidden, and with men who never boasted of their success with women. In fact, they were more attracted by male companionship and male frankness in speech than by the amorous experiences themselves. Most interesting of all—especially for those times, which were not like earlier times—no one particularly begrudged them their adventures. People knew, but kept still, as if by common consent. These women were generally widows, and from prosperous homes at that, women who were still strong, but with nothing to do, resentful of life and the unhappiness it

brought them. Evidently they formed the same stratum in
female society that the derelict officers formed among the
males. The men differed markedly from one another in
appearance, bearing, and manner, but these women were
even built like one another. Perhaps this was because they
belonged to upper society and had been reared especially
for the new officer caste under Prince Nikola. Or perhaps
the same mode of life left its common stamp. They were
all rather tall, feline in their movements, and without the
angularity so characteristic of Montenegrin women. All
without exception wore clean clothes. They put on veils
and shoes whenever they went out of the house or whenever they expected company. They washed with per-
fumed soap and secretly used pomades. They were not
ashamed to sit with men in the coffeehouses or to go off
with them on long trips.

One of them, Darinka Grujičić, was from my part of the
country and a close friend of my family. Her husband had
been an officer who was killed in the war. Her two children went their own ways, though she gave of herself to
them unsparingly. Left completely alone, constantly worried financially about hanging on to her property and
educating her sons, who did not care much for learning, she
lost time—and herself—in Belgrade.

Slim as she was, though already around forty, dark and
easy on the eye, Darinka seemed to want both to take her
husband's place among the officers (she was an excellent
horsewoman) and to be a woman. She never missed a
market day, not only for financial and similar reasons, but
also to be with the men, to show herself, and to get something out of life. She would gallop into town bold and
proud, as though she wished by her appearance to call attention to her renowned family and hero husband, whom
she mentioned with pleasure and pride, but rather too frequently. Her life, full of conscious pride and without joy,
consisted of riding on horseback from the village to the

town, of having a drink with the officers at the inn or at a
party, and of discreet and rare—and therefore all the less
satisfying—amorous adventures.

She rode a horse better than any riding master, as though
she found some special enjoyment in this, and spared
neither her mount nor herself. She was moderate in both
food and drink, though she emptied her glass like a man,
quickly and at once. She did not stand out in dress; she was
always in a black, though rich, outfit. She was very jealous
of her good name and reputation and was careful not to
bring shame on her dead husband. She would not even consider marrying a second time. Whom? A peasant? No, she
could never get used to the rigors of such a life. A burgher?
For her this was an alien and impotent breed. The officers
were all either married long ago or dead. She had to live
out that bit of life left to her as fully and yet as honorably
as she could.

This was the problem of thousands of peasant widows.
They lived a similar life, though in a different way—without such powerful and obvious desires, but curbing themselves in a life of toil. Darinka, on the other hand, had many
advantages. She did not have to work in the fields and
could afford to hire a servant. She was different from
ordinary peasant widows.

Going from my village to town on holidays, I would go
with Darinka. She would put me behind her on her horse
and I would clasp my arms around her waist. When we
went uphill, I could feel her soft sides gently heaving in
and out.

Darinka liked best of all to be in the company of her
husband's comrades. She seemed in every way to feel freer
with them. Thus she came to our house as well, but rarely,
and always with my father. She treated my mother with
special respect and consideration, and my mother paid her
in kind with a noticeable cordiality. On one occasion, while
my father was escorting her, her arm under his, she plain-

tively called him by a nickname. Was there something between them, that mysterious something that happens
between a man and a woman? Or was this simply her
tenderness for a beloved comrade of her husband's? Was
this in remembrance of her husband, of a carefree past filled
with pleasures?

Amid this society, which disappeared as its members devoured one another, she nevertheless stood out for her
perseverance and defiance of life, despite her awareness of
its necessities and of the senselessness of waging a struggle
against it. In all this there was something both sad and
powerful. However, awareness and perseverance are not
enough to help one resist and survive if the times in which
one lives are contrary to those that are ahead. A man can
fight everything except his own times.

My father's life, too, was incurably wounded. He, too, could not fit in or find himself.

Even after he was pensioned he was forced to be constantly alert for fear of ambushes. He expected the Rovči guerrillas to wreak their vengeance on him. The entire family lived in dreadful anticipation for two or three years. Our concern was all the greater because the guerrillas, whose numbers were dwindling and who were increasingly hard pressed by both the officials and the people, were becoming increasingly brutal.

Father traveled frequently and usually returned at night. The snorting of his horse would rouse us from a fitful slumber, but it was a joyful sign that he had once more come home alive.

When our dog Garov, who was on a leash, barked at night, we could tell whether it was at wild animals or at a man, for if the latter the barking was more constant and huskier. Awakened, we listened to sounds outside the house, which stood alone in the night and in the forest. Garov's quickened barking and jumping made clear that unknown and uninvited humans were approaching, and Father, who had waited in silence, took his rifle and shot out the window, to encourage the dog and to serve notice that he was awake and watching. Later, in the bushes by the little bridge over the stream, we found the grass had been trampled down, and there were still cigarette butts. Men had lain there in ambush; the earth was still warm from their bodies. Suspecting an ambush, Father avoided the bridge, and crossed the river downstream from rock to rock.

One summer night, as we were bedding down the cattle, some fifteen armed men popped out of the forest and descended on our house. My brother hurried home to tell Father, and behind him ran Todor Dulović's most devoted friend and helper, Radojica Orović, a slight youth but live as fire, ruddy, black-haired. Father rushed out to the steps with revolver in hand, but Radojica called out to him not to be afraid and not to shoot. Father was now on good terms with Todor, though not with the infamous Rovči guerrilla Milovan Bulatović. Now, apparently, Todor hoped to make peace between Father and Milovan. That very night Todor, Milovan, about fifteen guerrillas, and my father met in our house for a long tortuous parley.

Milovan Bulatović was a rather fair-haired youth with coarse features, huge head, wide mouth, and large teeth set far apart. He was crudely built, even by peasant standards, but quite a hunk of man. He spoke little and evasively. His smiles were rare and forced. It was said that he was treacherous and cruel, and cruel he certainly was. He was a man of great and savage bravery and endurance.

With his revolver in his pocket, Father presented his case as the guerrillas, mouthpieces for the silent Milovan, put questions to him. Father was neither frightened nor confused; he acted like a man ready for anything. It was apparent that he wanted to settle everything in peace. At times the parley hovered and hovered over that thin line which separates violence and reconciliation. The words would smell of powder and blood, and then in a few moments gain the human warmth of cordiality. Though we were children, my brother and I planned what we would do if Father were attacked. My brother intended to grab somebody's rifle, and I . . . Todor intervened, though rarely, making it seem that occasionally he was on Milovan's side.

He was an honest and realistic man. Had he been convinced that Father deserved death, he would have been the first to lift his gun. Reconciliation came late in the night. No one had offered the guerrillas any food until then; Father shouted, and soon mounds of beef appeared and brandy was spilled left and right.

This reconciliation, however, did not give Father any peace. It was then that the invisible reins of caution within him were loosed and he began to be goaded by a restless force that drove him to dissipate wildly both his strength and ours. First he undertook the job of digging irrigation ditches. He made measurements, drove in stakes, connected them with guidelines, and went in deeper and deeper, dynamiting rocks and excavating. Then the Tara spilled over its banks and, wild as it was, flooded and destroyed everything. He began to clear the forest in a steep unfertile place. This, too, was work in vain. He tried his hand at being a merchant. And so on—first this and then that. He even drank at times, though he never became completely drunk. He would come home at night tipsy, and, in a gay mood, would hand money around. In the morning, sober, he would take it all back, remembering exactly what he had given to each of us.

It was at this time he began his friendship with the priest Father Aleksa, who was much older than he, a man who was unusual in every way. The priest's only son had been killed by the guerrillas. It was said that he had been a little touched in the head ever since. Maybe it was true, but he was in every way an eccentric man even without this. The parish was quite scattered, and Father Aleksa was its only priest. He was always on the road, on horseback, and hardly ever in church. He visited the people constantly. Dawn found him wherever night had caught him. He generally picked at night the more prosperous houses, with softer beds and richer fare, and in his crude frankness he never concealed it. "Not even a votive light burns without oil," he would say. "Hard beds for the young, and young lambs for the old."

Father Aleksa was known especially for three traits: he spoke in short proverbs, which he himself invented; he was extremely frank to everyone; and he loved to drink brandy.

He was not one of those drunkards who guzzle themselves into unconsciousness. True, he never concerned himself with moderation. "Brandy is not a scale," he would say. Yet liquor never made him lose his senses, certainly never enough to topple him from his horse. God preserves drunkards, to cite his own words, so he could travel on horseback while drunk, in the foulest weather and on the worst roads. To tell the truth, he was not very steady when he was sober, if that he ever was. But his brown mare—a gentle and powerful nag—knew the roads and inns very well. Since the priest was accustomed to staying in only the best homes, the mare followed the same procedure, paying little attention to whether a particular time or house was the wish of the priest or the host. And so Father Aleksa would suddenly appear, old and bony, a huge head swathed in a hoary beard and hair—a real rock in a snow-laden bush.

Father Aleksa did not care much either for church or for prayer. As it was, he was in church but rarely, and he conducted all his baptisms and requiems in an abbreviated form, using only those prayers he held to be the most important. He carried a prayer book in his saddlebag, but no one ever saw him read it. He regarded it more as a holy relic than anything else, like a cross; he would place it on the table or altar and then chant by heart and very loud—then even louder and more distinctly, that God might hear better.

At that time my father, too, grew a beard, which he kept to the end of his life. He never told anyone the real reason for this, if indeed he knew. Maybe he just wanted to be different. Perhaps he wished to look more pious. At any rate, it certainly had something to do with his friendship for Father Aleksa. Unlike the priest, he combed his beard in halves and curled up each side. Father Aleksa told him that he now had four mustaches, and Father liked this very much. And because Father knew how to sing the responses in church, and very well at that, the priest told him that he could become a monk any time.

Father had no such intentions. He had not cared much for church, never went to communion, prayed only on his patron saint's day and on Christmas, and then perfunctorily and loudly, more to satisfy custom and as an example to the children than out of any inner conviction. But then Father began to get religious. He went to church, crossed himself every evening, and talked about the importance of a clear conscience.

Nevertheless, it was evident that Father Aleksa was not the one to turn Father to piety. Their friendship derived more from the circumstance that this strange and eccentric man railed, half in jest and half in earnest, against both of the principal political trends—the Greens and the Whites. He was a very lively opponent of both sides, without being for either. He had no program of his own. He attacked anybody in authority and power, and that was enough for him. He did not do so with malice or passion, but with a mock seriousness. He admitted about himself, "South wind up the Tara, north wind down the Tara." Actually, these winds blow in just the opposite directions, and by this he wished to say how everything about him was contrary to what it was with other men.

It is difficult to explain why, after all, this strange man was so beloved and respected by the people. Was it because he was a man of his word? Or because of his proverbs? Or unusual life? Or because the men of that time were attracted by eccentricity? What really drew Father into this friendship with Father Aleksa and into religiosity? Was it because he agreed with the priest in his dissatisfac-

tion with the political situation? Or because he felt some
burden pressing on his soul? Or because he was wandering
anyway? Or did he desire this friendship as another way of
being different?

Father and the priest never had a spat. This was all the
more unusual because Father Aleksa, too, was a difficult
man. Their friendship did not cool, but was cut short by
the priest's death. For once the faithful horse did not bring
its master home. The old priest fell drunk on the road and
there he remained. Father did not mourn him much. The
priest had already grown old, and Father had begun to get
over his religious mood. He could find nothing of permanent value in it. He had simply taken a walk with his good
friend Father Aleksa, seen how it was, and continued on
his own way.

Father's wanderings and enthusiasms became entangled,
for a time, with the misfortune of Uncle Teofil—Tofil, as
we called him. As a prisoner of war, Tofil had been sent
by the Austrian authorities to a labor camp in the Tyrol
and had fled from there to Italy. The Montenegrin government found him and settled him and other refugees in
Gaeta, the center for Montenegrin refugees who supported
King Nikola. Though a peaceful man, alien to politics,
Tofil, finding himself where he was, joined those who were
for Montenegro as an independent state. Time passed, however, and so did any prospect of an independent Montenegro. Tofil went back home. He found a devastated
homestead—his two sons and wife dead of typhus, and his
youngest son gone off into the world. Even before his return he lived in the tales told about him as though he had
never been away.

This man of marked features, with a yellow mustache,
blue eyes, and a perpetually pouting lower lip, was of a
very gentle and tolerant nature. He rarely got angry—only
when he had to, and then seriously. Stocky and taciturn,

he seemed rather stupid and rough hewn. Actually, he was extremely bright and possessed a deep and firm understanding.

It was not this that brought him renown, but his heroism, heroism of the rarest kind. He was not the sort of hero who boasted; he regarded his own heroism as something quite natural, something that came of itself, just as men wear clothes or drink water. He was beyond any doubt a soldier who had no equal, not only in the regiment, but in the whole Kolašin brigade. He distinguished himself in the Battle of Mojkovac, in which fortunes changed every hour. Strong, swift, and brave, and a hunter, besides, who had hunted rabbits with an army rifle and who knew the countryside very well indeed, he slipped behind enemy lines two or three times to collect data and to raid pillboxes and staff quarters. He would spend the whole night pushing through snowbanks and past enemy patrols, and appear with the dawn behind enemy lines to sow death and destruction, and the next morning, if need be, he would be back again. On one occasion he and a comrade wiped out a whole platoon of Austrians trapped in the snow and the forest.

This fearless hero, who had faced enemy soldiers and wild beasts, this fugitive from a prisoner-of-war camp and an *émigré*, returned finally in his long green overcoat to his razed hearth. Relatives took him in. He stayed with them a while as a guest, and then returned to his home, determined to rebuild and to start a new life. He was already about forty-five years old. That very spring he could be seen barefoot behind a plow drawn by borrowed oxen or clearing the plum orchard around the house.

The authorities regarded him with suspicion. Suspecting that he had returned with concealed motives, they called him into the district police station. Sergeant Adžić, who was known as an inconsiderate and stubborn man skilled in tracking down guerrillas and their supporters among the

Greens, slapped Tofil around, trampled him underfoot, and abused him in foul language. Then they let him go, for he was not guilty of anything. But the deep wound remained. This was the kind of affront that every man of honor had to wash away in blood, especially such a man as he. Tofil found himself in a dilemma—whether to avenge himself or to suffer the insult and devote himself to building a new life.

Here my father, himself discontented, interfered. He began to persuade Tofil to join the guerrillas. Father even offered to go with him into the forest. Quite possibly Father's desire to become an outlaw was not very deep; having once decided, however, he would have gone, come what may. Apart from all else, this was not a good time to consider such a course; the guerrillas were on the wane.

Father's mind was more penetrating and quicker, but also shallower. Tofil's slower but better-grounded mind found itself in conflict. His injured honor nudged him to my father's side and tempted him to a hasty decision. He struggled hard with himself. He wept late into the night, as Father angrily paced the floor denouncing the government and everything else.

Finally Tofil's real nature won out—he would rather build a life anew than destroy it. That deeply rooted drive for survival prevailed, the desire to re-create a life inherited from his forebears out of their primeval struggle with cruel nature and with clans of another faith. Montenegrin officers, my father among them, and perhaps Montenegrins in general, rather easily subordinated that irrepressible desire for perpetuating life and oneself in it to wounded honor and heroism. Tofil, too, had honor and heroism, even more than any of them, but that other drive in him was always stronger.

By that autumn he had taken a bride, not a very pretty one, but young and healthy. He begot two children with her. He made out well with stubborn effort; his property and children grew noticeably with each year. He built a new house, the finest in the whole district. In a few years he had become a prosperous and model homesteader.

The strongest are those who renounce their own times and become a living part of those yet to come. The strongest, and the rarest.

One person, however, did not change, as though neither time nor space existed for her. This was my mother.

She devoted her whole being to that which she regarded as natural and inevitable—begetting and rearing children, being good to her husband, and working slavishly. Apart from that no joy or thought existed for her. Calm, patient, unyielding, she seemed neither bright nor energetic. This was, in fact, not so. She accomplished everything by dint of tireless persistence; her mind was hidden away somewhere deep, perhaps in those bygone ages from which she, through her forebears, gained her simple and solid experience.

Nothing could surprise or frighten her very much. She mourned simply and deeply, without many words or much outcry. She rejoiced in the same way, unnoticeably, the way she breathed. She bore within herself certain immutable and strict rules of honesty, justice, truth, faithfulness, mercy, reliability. Though hardly visible and unproclaimed, these immutable laws burst forth in her instantly and with unusual vigor if anyone violated them. The foundations of her mind rested on these rules of decency, firmly entrenched and unshakable. Her mind was infallible. Father rushed from one idea to another, while she remained constant. In every moment of crisis for the family or for Father, her constant soul and simple and inexhaustible wisdom came forth.

In his youth, before his marriage, Father had had an

illegitimate son, whose mother was not reputable enough for my father, an officer, to marry. He was prevented in this, to be sure, by the mores of that time and place—an officer could marry only a girl of good family—and Father submitted. The child had been born while Father was still in prison under suspicion of seeking vengeance against the Prince for his father's death.

Even after his release, probably because of shame, Father never went to see the mother and her child. Montenegrin laws were very severe in these cases, and public opinion was particularly hard on unwed mothers. They were held in contempt even if they were later married to the father of their child. Father found it best to extricate himself from the whole affair; his imprisonment was an advantage. It enabled him to marry the one he wished to be his wife.

Everyone knew about that son of my father's, though all had been long forgotten. Mother, too, knew about this event in Father's life, but she put it out of her mind. Thus twenty or more years went by. We children knew nothing. Something happened, however, so that we, too, had to be told.

This son of my father's was very capable, as is frequently the case with orphans. He went through school by his own efforts, and even supported his mother, who had become mentally ill. The time came when it was very important for him, a grown man, to settle the question of his paternity. His reputation was hurt by the fact that he was an illegitimate child. The law denied him many rights that legitimate children enjoyed. With the help of reputable men and Father's friends, he undertook to have Father recognize him legally as his son, though he sought no share of the property or anything of the sort. Father agreed, which was not an easy thing for him to do, especially because of his family. Father was forced by circumstance to tell us all and to acquaint us with his forgotten son.

Father and son arrived one late afternoon on horseback.

The new son was one of those slightly built people who give the impression of being wiry and resilient. Mother greeted him first and deliberately embraced him. We all were in tears over Mother's generosity, which Father misunderstood as our joy on meeting our unknown brother.

Father acted like a father who had not seen his son for a long time. Once Father had decided to recognize him, parental feeling seemed to have welled up in him. Perhaps for every Montenegrin of the older generation it would have been the same. Anyway, that is how it was. Father conversed with his newly recognized son, showed concern over his problems, gave him unrealistic advice, and was even proud that a son of his should have succeeded in making his way in life. This feeling was slowly transferred to us, and what was most interesting of all, the new son began to act, in this atmosphere, as though he had at last gained a family, even though it must have been clear to him that there could never be any real natural ties, for these are created by living together. He had no reason to feel particularly grateful to Father; on the contrary, he had no little reason to feel resentful.

The new son departed in good humor, more because of Mother than because of any of the rest of us. She gave him a warm send-off, packed him a lunch for his trip, and gave him beautiful woolen stockings. He hardly needed them in the city, but he was unusually touched. Through that infallible inner moral sense of hers, Mother understood how important it was to help a man who was suffering through no fault of his own, and how everything else was immaterial, even her own feelings. Without any inner struggles whatsoever, she received this illegitimate child of her husband's.

But a friend of hers, a woman of the same character, was nearly destroyed by a family crisis of her own.

Jaglika, from my mother's part of the country, was married to a Montenegrin officer named Marko Petrović.

They, too, had male children, already quite grown, when a sinister and cruel drama began to envelop them.

There lived in their house, as a servant, a young Moslem, illiterate but rather bright and energetic. Marko's sons grew close to this good-natured lad, helped him in his work, and even taught him to write. Jaglika treated him, a servant, as a member of the family at meals and in every other way. The youth felt quite at home.

Marko, who was older than his wife, aged rapidly while Jaglika remained unchanged. As he grew old, Marko felt an unquenchable passion, which became all the fiercer as it began to wane rapidly. Perhaps he had formed an attachment with a younger woman, wanted an excuse to get rid of his wife, and, since he could not find one, hatched an infernal and foul plot. To cast suspicion on Jaglika, he forced her to have a talk with the youth by the fireplace and to get him to reveal their alleged liaison. Marko, pistol in hand, was to listen in the attic.

This was a terrible shock for Jaglika; her husband had never suspected her or ever had reason to. She was known throughout the countryside as an honest and completely upright woman. She was torn by a deep anguish such as she had never known. What would people say about her, the mother of grown children and a woman already well along in years, if her husband failed to be convinced of her innocence and started a scandal for which she was not to blame? What would her children say? What did her husband want of her?

She agreed to that meeting with the servant and began by expressing to him, rather awkwardly, her supposed desire. Unable to continue she began to inquire his feelings for her. The young man could not comprehend at first. When he did, he began to sob, saying that he had always regarded her as his mother. Jaglika, too, began to weep. Marko did not appear from the attic.

Though brought to shame, Marko still could not free

himself of this wild and malicious imagining. He told Jaglika that she and the servant were merely pretending, that she had given him some sign that Marko was listening. The tortured woman could stand it no longer and confided in her oldest son, already a young man, and he told his younger brother. What terrible thing had been spawned in the soul of their father, who had been tender toward his children, a good husband, respected by the people, and a hero in the war?

The sons took their mother's side, determined to quit the house if their father did not change his attitude. But Jaglika prevented any encounter between father and sons. She let her husband know that, in desperation, she would tell all to their sons. This sobered him. He withdrew quietly. He even became good to the servant, and especially so to his wife. He bought her some presents in town. It all ended well. Marko's relations with his wife changed. She found an even stronger and more independent position in her marriage. Growing old and weak more quickly than Jaglika, Marko came to depend on her more and more, and his occasional outbursts of temper became all the more foolish and comical, even to himself. He was no longer the rough, unbending, and often arbitrary master of the house.

Something similar happened between my father and mother. As he neared the end of his life, he became more and more gentle, yielding and tolerant in everything, while she fought all the harder for every little thing that could secure a better future for her children and house.

Folk songs and epic poems give the impression that manliness, heroism, and a simple purity were the exclusive qualities in the life of the people. This is not so. Once, long ago, life may have been like that, or we remember only its virtues. But life today is not. There is much that is murky and unfathomable in the soul of our people —simple, heroic, poor, and honorable as they are.

Constant wars, burning, massacres, plundering, despotism, and exploitation, every kind of violence and cruelty, all went on and on and begot manliness and heroism, but also many other things.

Man is not simple, even when he is of one piece. That piece has many corners and sides.

The guerrilla movement fell off rapidly, leaving behind bloody and foul traces, as though the guerrillas, who were dying heroically and sinking constantly to lower depths of violence, themselves wished to besmirch what they had fought for.

The causes for this decline were manifold. Suppressed by the authorities and tired of living a life under siege, the people forsook the guerrillas more and more. Embittered, the guerrillas in turn committed acts that separated them from the people, that one forest thickest of all in which they might have found refuge.

Once they broke into the house of a peasant whom they believed, perhaps rightly, to be a dangerous spy. They fell to drinking and eating. And while they were feasting thus, they slowly turned him on a spit by the fireplace. The next day, still drunk and frenzied, they crept into a hidden glade and massacred some ten soldiers who were peaceably passing by.

Up to that time the guerrillas had avoided conflicts with the army. They clashed only with the gendarmes, arguing that soldiers were recruits and not to blame if the authorities sent them on raids. But both these incidents had a disturbing effect on the people, and were incidents that the authorities were able to exploit easily. Events like this took place not only in our district, but elsewhere in Montenegro. Sensing their defeat, the rebels became increasingly impatient and embittered.

Todor Dulović, however, remained the same.

He came to Aunt Draguna's, where he felt quite secure. It would have been difficult to imagine that he would come so near the city otherwise. He would have come even more frequently had not Draguna and her husband feared their son, Ilija, an implacable enemy of the guerrillas.

Fear that something might happen to Ilija was the most important, though perhaps unconscious, reason why Draguna's family treated Todor so cordially. After Ilija left for his new post in Serbia, the family began to dread Todor's rare visits. Milosav even had words with him, rather sharp ones, blaming him for not preventing the murder of his relative Arsenije, though Todor swore he knew nothing about its planning. Todor's oath on that occasion was a strange one, as though he expected another kind of death. He said: "I know nothing about that death. May Arsenije's wounds never befall me."

The winter of 1921–1922 was a hard one for Todor. His hideaway was discovered and he and his men driven into the frost and snow. Pursued, they treaded the water upstream, in all that cold, hoping to cover up their tracks. The odds were against them, and finally they were compelled to split up to keep the band from being taken as a whole. Todor remained completely alone, wandering over bare and snow-covered Sinjajevina Mountain in midwinter.

This was the first time, he later said, that he had taken food by force. He came across a cabin one day and asked the shepherd for bread. The shepherd refused. He did not have a single crumb, he said. Todor looked, and on a shelf a newly baked slab of corn bread was steaming. The shepherd knew Todor never took food unless it was voluntarily given. Perhaps, too, he had grown bold toward these rare and hounded rebels. But this time Todor did not go hungry; he took the bread.

Todor also killed a man, against his own will, during that winter. A certain peasant set out to track Todor down on the bare mountain. The peasant was rested, agile, with food in his belly; the guerrilla was hungry and chilled. Todor saw that the peasant would overpower him if he reached him, and yet Todor could not escape. He begged, he entreated the peasant to turn back. But the peasant would only stop a while, then continue to follow the guerrilla. The peasant could see that Todor barely had breath in his body; he bided his time until Todor fell exhausted in a snowbank. He wanted the guerrilla alive. He wanted the reward, and the glory. There was no other way out; Todor raised his rifle—and killed him.

Todor now looked worn out, though he would not admit that he was. He suffered from frostbite after tramping through the snow. His feet were gangrenous and exuded a heavy odor. Yet he suffered without complaining.

When Todor's second brother, Mihailo, lost his life, a sheaf of poems was found on him. The young man had never participated in the fighting, but wandered through the villages, from widow to widow, bewailing in verses his lost youth and guerrilla defeats. Though quiet and calm, Todor, too, felt some sort of powerless sadness. He carried with him always an old, tattered newspaper. In it was a poem in which King Nikola, not without pathos, reproached those of his men who had deserted him.

Todor always said that he was fighting for an ideal. What is an ideal? Something altruistic, like Todor? Something good which men can be told and which they will store in their heads, just as Todor did?

Todor's faithful comrade Radojica Orović was quite different. He gave no evidence of suffering. He was younger and as tough as a knot. Todor was quiet and sober, stable and sure in everything, like an ancient epic poem. Radojica was vibrant, venturesome, and wispish, like a peasant love song. In Radojica one could see everything—shrewdness and bravery and wiliness. In Todor one could not see anything; no one could tell that he was

so great, manly, and hardy a hero. Here was a man who for
years lived with his gun the way others live with the plow.
Yet these two men got along well, perhaps precisely because they were different.

Todor lasted out the winter. But the next spring, 1923,
he did not go back to the green forest. He, too, lost his
life. On the way to the Vasojevići to avenge a blood
brother, he spent the night in the house of his blood
brother's brother, who had called him to seek vengeance.
This very man betrayed him to the police and placed a
powder in his drink. Todor fell asleep, drugged, in the
living room. Radojica Orović slept in the front room. The
gendarmes did not even notice him as they fired at Todor.
The first volley shattered the lamp. Radojica sprang into
the darkness, but they caught him alive in the melee. Then
the gendarmes rushed in. Riddled with bullets, Todor was
still alive, but unconscious. They fired into him and struck
him with bayonets. But he, they say, would not die. So
they cut him to bits. There was hardly enough left to
bury.

Radojica was later sentenced to twenty years of hard
labor. All the murders were placed on dead Todor's head,
and Radojica was tried simply as an accomplice.

Todor Dulović was the last renowned guerrilla in our
part of the country. Nor were there others of note elsewhere in Montenegro. Other guerrillas were held in less
regard than Todor. Most of them were killed, though some
were caught alive. Among the latter was Milovan Bulatović.
He was surrounded, and surrendered after his band was
wiped out. Though the people were sick and tired of the
guerrillas, they did not approve of their surrender. They
believed that they had chosen their path, and should have
traveled it to the end—death, not surrender.

But once in the hands of the authorities, Milovan gathered his strength again. He did not betray his partners
or confess to any murders. He held fast, even when he was

sentenced to death. His execution in Kolašin by a firing squad was attended by a multitude of people.

Slightly bent and pale, but with a calm and resolute demeanor, Milovan walked handcuffed between two gendarmes to a grave dug in the meadow near the town. An intolerable stillness reigned while the sentence was read, the last rites offered, and so on. It lasted far too long. It seemed that Milovan, more than anyone else, was in a hurry.

Nobody expected that a guerrilla who had been tortured by prolonged interrogation, fruitless waiting for an amnesty, and the lengthy ritual of that morning, would be capable of saying anything. However, as he was bound to the stake, by the grave, and as he looked straight at the row of muzzles—he had not allowed himself to be blindfolded—he was suddenly seized by a desire to take leave of his land. With all the force in his breast he shouted, "Farewell, Monte . . ." He never had a chance to finish. The volley cut him down in the middle of the word.

Immediately after Todor, there died another renowned guerrilla, Drago Preljević. Todor had been an idealistic rebellious peasant, but Drago went into the forest because of blood revenge and only later became a political malcontent. He was quicker to reach for his gun than Todor was. Still very young, under twenty-five, he was regarded as one of the most dangerous rebels, even though it was said that he showed a very gentle nature when he was not fighting.

Previously, he had lived peacefully in his village of the Bratonožići as a shepherd, while his older brother, Vučić, was a savage guerrilla. The powerful and renowned Orović family of Lijeva Rijeka, who supported the unification with Serbia, took part in Vučić's murder. Bitter, Drago left his cattle to take up his gun, determined to seek vengeance and to wipe out the Orovići to the last man. He joined the guerrillas in the forest and took part in clashes with the gendarmes, and in many murders. Usually, though, he was a lone wolf. The others were not interested in wiping out the Orovići, but he had taken an oath.

Like Todor, Drago Preljević had a big bounty on his head—one hundred thousand dinars. He killed four men from the Orović clan alone. The number of others whom he murdered was significantly greater, according to a list found on him together with some sad folk poems in which he mourned his brother and boasted of how gloriously he had avenged him.

I saw him only once, and he was dead.

He was killed by treachery, while asleep near Kolašin, by Vule Lakićević of Rijeka Mušovića. Vule was an ashen man with pale eyes and a long nose, unnaturally long arms and legs, and almost hairless. He was very tough and hard; if anyone was a born killer, he was. He was accustomed to living off the blood of other men, and he killed other rebels after Drago. One day, when there were no more rebels, he became a hired killer who took money from those who wished revenge but dared not risk their own heads. Blood revenge became a crime, and those who carried it out became hired killers. Vule was involved in shady deals. He did no work, yet lived well. People avoided him, as though he were a living vampire. At long last, just before the last war, he landed on the gallows for murder—he had killed a young man for money. Vule went to the gallows coldly, as though this were the fulfillment of his life on this earth. Such was the man who had killed Drago Preljević, whom blood revenge had driven to terrible deeds.

Vule himself had previously been a rebel, but became a well-known informer. The authorities had bribed him to do away with Drago. When the news of Drago's end came, the authorities rushed from Kolašin to see the sight, and to make merry. In Smaïl-Aga's field above the town cemetery, a multitude gathered, not to pay homage to the dead bandit, but out of curiosity, to see his body as it passed by. Late in the afternoon came the officers and district chief, and with them the murderer. Vule carried two rifles—his own and Drago's, slung around his neck.

After the authorities and the murderer had left the scene, a group of people who had known Drago took over. He was dead, and there was no one else to care for him. He was now just a showpiece. They tied him to two poles, a makeshift stretcher, which were dragged along by a scrawny mountain pony. Drago's head, lower than his feet, jogged over the rough road. There was no peace even for the dead. The group stopped by a bare field near the cemetery and dumped the corpse there.

At that instant rain came pouring down, and the people scattered, leaving Drago alone on the field, a small man in a dirty and torn suit, with handsome features, blond curly hair, and a little beard. A bullet had crashed through his forehead and shattered his skull; a piece of his skull had been lost somewhere along the way.

When the rain died down, two gypsies appeared and started to dig a grave. Before they were through, the rain started again. They threw him into the wet and slimy grave, without a coffin or even a prayer. There was no one to utter the wish that the black earth might rest lightly upon him.

Two days later, his sister came to him. One could hear far in the distance how she wailed and keened over the solitary grave.

Sisters mourn, and are comforted. Who shall mourn the times and the terrible fate of men in them, a fate for which not they are to blame, but circumstances not of their choosing?

The Montenegrin King Nikola also died. Nothing was left of the once mighty sovereign who could break men like twigs and play with them like children. Time had out-witted and outstripped him, too. His sons were good for nothing, and his sons-in-law not only deserted, but turned against him. There remained, in the end, only a touching memory of an exile in a foreign land who died clutching in his hand a Montenegrin stone which he had taken with him when fleeing from the Austrians in 1916.

Neither King Nikola's death nor the end of Todor and other guerrillas surprised or upset anyone. All knew this would come. Their time was past.

The murder of the more notorious guerrillas, like Todor Dulović, made it easier both for the authorities and for those with whom the rebels had accounts to settle. Such was the case of Bošković.

Boško Bošković was the most renowned personality in the whole district at one time. He best expressed the traits of a Montenegrin chieftain who had preserved the traditional virtues and yet managed to adapt to the new circumstances. The son of a famous rebel leader from the Tara Valley, he belonged to the younger generation of Montenegrin officers, and did not conspicuously stand against the despotic rule of King Nikola. He was a great hero in the wars, extremely severe with his soldiers, especially with the cowards and pilferers, whom he slapped and pulled by the nose before the whole company. When, in the prisoner-of-war camp at Boldogosony, he had to declare himself for or against unification with Serbia, Boško was at first resolutely against an independent Montenegro.

Such men, tough Montenegrin officers who were supporters of the unification, were the best fitted to carry out Montenegro's merger with Serbia. They were products of the Montenegrin environment, and their ways and methods of government were already familiar and sanctioned by tradition. As such, they could have accomplished more than a Serbian or any other outsider, even if he were more capable, for his attempts would be resented as something alien and unnatural.

The establishment, survival, and destruction of political systems in this land does not take place without great violence. Behind unification stood Belgrade and Serbian force. But the deeds of violence were carried out by the sons of these mountains, in whom savagery and violence were inborn and without measure. Though the people dubbed the supporters of the unification "house burners," still they seemed to understand that their acts of violence were inevitable, almost natural under the conditions.

Boško Bošković was just the man to put a quick end to resistance and to enforce authority and bare compulsion upon the people. There were others—politicians and wise-acres—who accomplished this more slowly through shrewd persuasion. He was not one of these. There were those who knew only how to carry out the orders of others. brutally and without much thought. He was not one of these either. He was quick and penetrating as well as ruthless in carrying out his purpose. He had a tough and unyielding, though transparent, will.

Boško's very body seemed to be built for such a cruel and violent task—to break his opponents. He was a rather stocky man, black-haired, and with a big head, stout and strong limbs, and a broad chest. There was something extremely forceful and arrogant about his carriage, his glance. One could feel, from the very first encounter with him, that he was heavy handed and thickheaded.

Once the task was accomplished and the opponents of the unification were crushed, Boško was left jobless and became superfluous even to himself. He left Kolašin for another post, but did not seem to care for it. He, too, was changing and becoming lost.

He possessed three great Montenegrin qualities—bravery, loquacity, and hospitality. But the greatest of the Montenegrin virtues—manliness—was apparently not his strongest point. He was not reticent about choosing the means to a given end, especially if breaking resistance to authority was involved. He was not born to rule—this requires more flexibility—but he could beat a path so that someone more skillful and shrewder could come on his heels to rule. His personal qualities at that time were not enough to let him play such a role and become somebody.

He grew richer by rapidly grabbing up Moslem lands in the Sandžak, something that never occurred to him in Kolašin. True, he was not very prosperous. But his wife Neda was wealthy. She constantly helped him financially, so that he did not have to worry about material necessities, an inevitable problem for other half-peasant Montenegrin officers. Though a big spender, Boško was not a spend-thrift. He particularly liked expensive and handsome suits, and on market days would change his clothes as often as three times. This was incomprehensible in that Montenegrin environment, yet nobody held it against him. Somehow dandyism seemed to go with his forceful appearance, to enhance it.

It was bruited about that he was too fond of women. To be sure, there was some truth to this. But it was even truer that they were mad about him, they adored him. He radiated opulence, masculinity, and bravery, and intoxi-cated them. His personality roused the imagination of even the most virtuous. They found good reasons to justify his weak points. He was still young, handsome, rich, never did a day's work. Whenever the conversation turned to his amorous exploits, these virtuous woman defended him and attacked his paramours. To put their argument in blunt words and metaphor, even a dog is quiet until the bitch wags her tail.

Other officers, as well, liked women and played at being lovers, but Boško was a very special case. His reputation and relentless violence made him all the more attractive. And while, for woman-chasing, others lost their good name, he seemed to gain repute for the very same thing. It was a part of his make-up. No one seemed surprised that even his wife was resigned to his love escapades. It was said that she knew all about Boško's high life but did not mind very much. It was regarded as natural and sensible of her.

Neda was truly a lady of rare finesse and intelligence. She was from an old merchant family, the Kalištani, of Podgorica. Having no brothers, she had inherited all the houses and stores. Her regular income paid Boško's expenses. She left Podgorica only on exceptional occasions, for she was busy managing her property and rearing her two sons. She visited her husband rarely, but always received him tenderly and joyfully. She shared that feminine philosophy, let the husband stray, as long as he comes home. But there was something else. She, too, saw in him what all the rest saw. She liked that about him which she could not find either in her town or in her family—the freshness of mountain blood, the brave demeanor of a warrior, and the strength of a complete male. The tame, soft, mercantile nature of the Zeta Valley seemed to seek the wildness and toughness of the mountains. Neda Kalištanka, as she was called, enjoyed great respect, which was increased by her tolerance for such a husband, a born he-man whose vices suited him.

In settling accounts with his enemies, Boško was ruthless, yet he was not the kind to invent sophisticated cruelties. The whip, the fist, and the boot were the only means he used, and he used them personally, not assigning the job to his underlings, though he did not go out of his way to curb the worst of them. Their behavior was accepted, in that time and place, with tacit disapproval but as something inevitable, and therefore understandable, something to be quickly forgotten.

One thing he did, however, was never forgiven, or forgotten. It was rumored that in his relations with women he misused his power. Suspicious young wenches, or any pretty woman without anyone to protect her, were brought to his office. The guards would stand watch at his door, keeping everyone out. One girl of the Rovči clan put up a fight, and her screams could be heard outside. Enraged, Boško beat her with his whip so that her blouse was all torn and her skin underneath was in shreds. There were others, too, who resisted, but this girl from the Rovči was like Boško himself; she fought back with all her strength and fury.

While my father was still a commandant in the gendarmery in Kolašin, Boško was wounded in one of his amorous exploits.

There lived in Kolašin at the time a shoemaker named Rakočević. He had an extremely pretty and hot-blooded little wife, all curves and softness, with a gold tooth which gave her smile a particularly fiery and devilish gleam.

Although he was an artisan, Rakočević had the reputation of being a brave man. Once, he surprised Boško with his wife, and as Boško was making his exit through the yard, Rakočević nicked him with a bullet—merely a flesh wound. Furious, Boško dragged Rakočević into district headquarters with the help of some gendarmes and beat him to a pulp. A real scandal ensued, all the greater because the event occurred during the Rovči rebellion against the authorities, while Rakočević was known as a fiery supporter of the unification. Boško was not bed-ridden, yet the affair could not be concealed. The aroused town forced the police immediately to release the shoemaker, mottled and swollen from his beating. To hush up the scandal, the authorities spread their own version—that Boško and Rakočević had had a quarrel, which came to blows; Boško allegedly wished to hit the shoemaker with the butt of his rifle, but the gun was loaded, and while he was defending himself, he pulled the trigger accidentally. Of course, my father, also, participated in spreading this legend. As he told it he would smile deceptively and with an obviously knowing air. Everybody knew the truth, and yet this tale was accepted, for it saved face for everyone, including the authorities. In this way the shoemaker and his wife were reconciled, and even Boško and the shoemaker. Perhaps the same principles, the same strategy, acted to curb both Rakočević's honor and Bošković's rage.

My father and Boško Bošković were friends, but not the inseparable kind. They had been in the same service, even before the war, and did little favors for one another. In fact, they were a bit afraid of one another. My father disliked Boško's violent nature, and dreaded the day Boško might turn against him in a way that would leave him little choice. Boško, on the other hand, feared the same; Nikola Djilas, also, was not exactly slow in going for a gun.

Their friendship went on for years, without any cordiality. Nevertheless, the day came when the two had a conflict and harsh words in a coffeehouse in Kolašin. Boško held it against my father that he, who had never been resolutely for the unification, was now with the government party, the Radicals, while Boško was sincerely an Independent Democrat and a follower of Pribičević. After the unification, Svetozar Pribičević * had come out for integration and centralism and a strong hand in internal affairs. This stand of Pribičević's came closest to suiting Boško's role and conceptions.

With men who give much weight to words, harsh words are the same as wounding one another. The quarrel between Boško and Nikola did not flare up in the coffeehouse, but neither did it die down there. Both seemed to be biding their time. Then it happened that both were going to Mojkovac, and they decided to travel together and have it out between themselves. This was like deciding

---

* Svetozar Pribičević was a minister in the unified kingdom in the period from 1919–1926. Originally he favored "centralism" in Yugoslavia's government, but after 1926 he formed a democratic reform party that favored a greater autonomy for the provinces that formed the kingdom, and he effected close relations with the Croatian Peasant party of Radić.

to spill each other's blood. The quarrel broke out in earnest then, face to face. Both were on horseback and armed. When they had grown tired of exchanging insults, Boško provocatively asked Nikola, "Do you intend just to talk, or did your mother bear you for something else?" Nikola suggested they find a suitable place for a duel, draw their guns, ride at one another—and let the best man win. Boško agreed.

They galloped along in silence for quite a while, though already in a deserted region. Behind them lay comradeship in arms and internment together, the same battles, friendship, and a reputation as sensible men. Yet now they were about to murder one another.

No one knows how their reconciliation took place and who was the first to give in. Nikola later said that he was, and Boško claimed that he was the one. Perhaps it came about because there were no witnesses to look on and to incite them with their presence. They made up, and both returned to our house for the night.

Boško was then, in 1923, rather stouter than he had been in 1919. Stoutness is highly regarded in Montenegro, but only if it does not go beyond the point of strength and vitality. He had already reached that point.

A real feast was prepared at home on that occasion, and all the more distinguished villagers were invited. Soldiers and officers shared their war memories. There were parried jokes and anecdotes. Happy warriors made merry, but were somehow restrained, as though by the fear of starting the quarrel anew. Everything ended well and happily, as it should among men who have decided to forgive and forget.

Later, Boško's relatives arrived to greet him. They disapproved of his traveling alone, even though, especially after Todor Dulović's end, he would seem to be in no danger, at least not any more than the rest.

And yet, in the autumn of 1924, he was killed in an

ambush on the way from Šahovići to Mojkovac, on a commanding height of Cer Mountain, as though he himself had
chosen where to die. He was riding with a relative, a boy,
as an escort. Heavy and stout as he was, he fell after the
first volley, but did not die immediately. He called to his
relative to fire and not to allow them to dishonor his dead
body. He was afraid they would mutilate and cut him up.
He wanted to look his best, a man of strength and good
looks, even in death. He was dressed in a gorgeous Montenegrin costume. So they buried him as he was, without
changing his clothes, as on the field of battle, next to the
wall of the church of his clan.

His death produced a dire effect and even worse consequences.

Boško Bošković was the last chieftain of the Polje clan.
The Poljani realized that his death marked the end of the
last living trace of their history, the uprising, the long and
bitter border struggles, and the campaigns in the great
wars. The whole region came to the funeral, and many
from other parts of Montenegro. Honor was done not only
to a hero, but to a heroic district, to a family that had
achieved leadership by the sword, and to a clan that was
vanishing.

Suddenly everything was forgotten—the internecine
feuds, Boško's violent pacification of these very Poljani
in 1919, and all his weaknesses and faults. There remained
only his heroism and glorious name, which personified the
heart and soul of the clan.

The murder had taken place in the Sandžak, that is, on
the other side of the Tara—long a bloody border between
two creeds. Consequently, it was not difficult for the
mourners, the keeners, and the eulogists to incite the masses
to a punitive massacre against the Moslems.

Other circumstances, as well, contributed to such a campaign. It was most natural to suppose that Boško Bošković
had been killed by the notorious Moslem rebels Yusuf

Mehonjić and Hussein Bošković. Yusuf and Hussein were begs whose lands had been expropriated during the land reform. Yet this was apparently not the only reason for their outlawry. They could not abide the infiltration of the Montenegrins into their region and the rule of the cross over the crescent. The Moslem population encouraged them, and even the Orthodox had admiration for their courage. Usually they roved about in the summer, and in the winter they would cross into Albania, where Yusuf was eventually killed by a bullet paid for in Yugoslavia, while Hussein sought final refuge in Turkey.

What greater delight could there be for the avengers of the Prophet's faith than to waylay and kill a renowned Montenegrin chieftain? To the Montenegrin way of thinking this was as though all Moslems were to blame for Boško's death. On whom were they to take revenge? On two elusive bandits? Could not the other Moslems have prevented the two from killing? Men such as Boško Bošković are not murdered without a big plot. There were other, more concrete, reasons, which the masses did not even suspect. Some politicians wished to weaken the strength and unity of the Moslems, who were banding together after their misfortunes during the war. These men, too, incited the aroused people to rise up in a crusade against the Moslems.

The main reason, however, was in the people themselves —a centuries-old inborn hatred against the Turks, a desire for vengeance for what the Moslems had done in the recent past, and a spontaneous hunger for Turkish lands, which the Moslems had held unlawfully since the Battle of Kosovo in the fourteenth century. Nobody, perhaps, felt all of this clearly, but it was evident to all Montenegrins that Boško's death could not pass without a disaster for either the Montenegrins or the Moslems, and that a new life was impossible without a general settling of accounts.

It was not difficult in such an atmosphere to inflame

hatred and to suppress everything that was reasonable and noble in these mountaineers. They had shed their blood unsparingly in wars that had brought them nothing, and now . . .

Immediately after Boško's burial, without any special consultation, the Poljani, and others with them, took their concealed rifles, and marched on the Moslems. Half of them were unarmed, but weapons were not necessary. The Moslem population against whom they were marching was unarmed, and most were not warlike, except those who lived along the former border, the Tara, most of whom had moved farther into the interior in 1912 or after 1918. The Montenegrins were not particularly well organized. They placed themselves, quite spontaneously, under the command of former officers, now pensioned, whom they had brought along and urged into the lead.

Never was there such a campaign, nor could one even imagine that this was hidden in what is called the national soul. The plundering of 1918 was an innocent game by comparison with this. The majority of the crusaders were themselves later ashamed of what happened and what they had done. But—they did it. My father, too, who was not particularly given to cruelty, at least not more than any other Montenegrin, never liked to talk about it. He felt shame for taking part in those events, like a drunkard who sobers up after committing a crime.

The police officials in the little town across the Tara as well as the civil authorities in the communities were mostly Montenegrins, and in the hands of the aroused mobs. In Šahovići the authorities informed the vigilantes that a group of Moslems, taken under protective custody on the pretext that their lives were in danger, were being moved to Bijelo Polje. The Montenegrins lay in wait for them in a likely spot, and massacred them near the cemetery at Šahovići. Some fifty very prominent Moslems were killed. A similar attempt was made on the Moslems of Bijelo

Polje, a peaceful and industrious people. They, too, were
to be convoyed by way of Šahovići under a safe conduct.
However, at the last minute a Serbian army officer prevented the treachery and crime.

The destruction of Moslem settlements and massacring
of Moslems assumed such proportions and forms that the
army had to be sent to intervene; the police authorities were
passive and unreliable. The incident turned into a small-
scale religious war, but one in which only one side was
killed. If, as rumor later had it, Belgrade wished to exert
some pressure on the Moslem party, which is not very
likely, the whole affair certainly got out of hand. Neither
Belgrade nor the leaders of the mob could keep it in hand.

Despite all this, not everyone was massacred. Holding to
the tradition of their fathers, the mob killed only males
above ten years of age—or fifteen or eighteen, depending on
the mercy of the murderers. Some three hundred and fifty
souls were slaughtered, all in a terrible fashion. Amid the
looting and arson there was also rape, unheard of among
Montenegrins in earlier times.

As soon as the regular army appeared, the lawless mob
realized that the matter was serious and immediately withdrew. After that the Moslem villages slowly withered. The
Moslems of that region began to migrate to Turkey, selling
their lands for a trifle. The district of Šahovići, and in part,
also, Bijelo Polje, were emptied, partly as the result of the
massacre and partly from fear. The Moslems were replaced
by Montenegrin settlers.

The affair produced general horror, even among most of
those who had carried it out. My older brother and I were
shocked and horrified. We blamed Father for being one of
the leaders of the mob. He himself later used to say that he
had always imagined the raid was intended only to kill a
few Moslem chiefs. Expressing abhorrence at the crimes,
Father nevertheless saw in it all something that my brother
and I neither would nor could see—an inevitable war of

annihilation, begun long ago, between two faiths. Both
were fated to swim in blood, and only the stronger would
remain on top.

Although Yugoslavia at that time had a parliamentary
government, the whole crime was hushed up. Had anyone
conducted even the most superficial investigation, he might
have exposed those who had committed the crimes and
their leaders. But there was no investigation of any kind.
Two or three guards were given a light jail sentence in
Šahovići because they had agreed to hand over some
prisoners to the mob. A general investigation was announced, but it turned out to be a travesty of justice.

What especially upset the established mores was not so
much the murders themselves, but the way in which they
were carried out. After those prisoners in Šahovići were
mowed down, one of our villagers, Sekula, went from
corpse to corpse and severed the ligaments at their heels.
This is what is done in the village with oxen after they are
struck down by a blow of the ax, to keep them from getting
up again if they should revive. Some who went through
the pockets of the dead found bloody cubes of sugar there
and ate them. Babes were taken from the arms of mothers
and sisters and slaughtered before their eyes. These same
murderers later tried to justify themselves by saying that
they would not have cut their throats but only shot them
had their mothers and sisters not been there. The beards of
the Moslem religious leaders were torn out and crosses
were carved into their foreheads. In one village a group
was tied around a haystack with wire and fire set to it.
Some later observed that the flames of burning men are
purple.

One group attacked an isolated Moslem homestead.
They found the peasant skinning a lamb. They intended to
shoot him and burn down the house, but the skinning of the
lamb inspired them to hang the peasant by his heels on the
same plum tree. A skilled butcher split open the peasant's

head with an ax, but very carefully, so as not to harm the torso. Then he cut open the chest. The heart was still pulsating. The butcher plucked it out with his hand and threw it to a dog. Later it was said that the dog did not touch the heart because not even a dog would eat Turkish meat.

It may seem, if one reasons coldly, that it hardly matters, after all, how men are killed and what is done with their corpses. But it is not so. The very fact that they treated men like beasts, that they invented ways of killing, was the most horrible of all, that which cast a shadow on the murders and exposed the souls of the murderers to their lowest depths, to a bottomless darkness. In that land murders themselves are not particularly horrifying; they are too common for that. But the cruel and inhuman way in which these were committed and the lust that the murderers frequently felt while going about their business are what inspired horror and condemnation, even though Moslems were involved. True, there was an already established opinion that one religion must do evil to another, and man must do evil to man. There is the proverb: Man is a wolf to every other man. People seemed to believe that a man who does not act thus is not human. But these crimes surpassed everything that had come down from the past. It seemed as if men came to hate other humans as such, and that their religion was merely an excuse for that monstrous hatred. The times had unnoticeably become wicked, and the men with them. After all, it is the men who make the times.

As a final injustice, it was not Moslems who had killed Boško in the first place, but Montenegrins, chieftains from Kolašin. My father found this out later from a trusted friend. The chieftains were envious of Boško. Just as they had once feared his forcefulness and power, so now they envied his wealth and rising good fortune. There were also many unsettled accounts from bygone days. Carefully picked assassins waited for Boško six days and six nights on the same spot. It was known that he was going to attend a certain feast, but when he would set out was kept a secret. Finally their patience was rewarded. But the chieftains who had organized the assassination did not lift a finger to prevent the massacre of the Moslems. They understood that Boško's murder was only an excuse, and they rejoiced secretly that Boško and the Moslems were being wiped out at one fell swoop.

Such were the terrible consequences of the death of Boško Bošković, a man whose life had been full of grandeur and horror.

Sekula, who had cut the ligaments of the Moslems' heels, on the other hand, hated the Turks more out of an inborn urge than out of criminal tendencies. He, too, was dissatisfied with the political situation and detested the existing powers—the gendarmes. He felt unrewarded for his sacrifices and exploits in the wars. He had looted and, secretly, killed Moslems even earlier. He was not alone in this, and neither he nor the others felt the slightest twinge of conscience. Nevertheless, he stood apart for his cold hatred, of which he was proud. He exulted in it.

In constant difficulty with the police—investigations, feuds, and smuggling—he finally fell victim to those who could not tolerate human beings any more than he could stand Moslems. A group of gendarmes gave him such a beating that they ruined his body, and he died soon after.

He had actually been a fine figure of a man, and loved to treat others. He was of slight build, but all energy. Even his eyes were like that—they danced, black and fiery. He dressed neatly and was clean, like a man who never works, or who works but rarely. Though poor, he did not stint toward others and behaved with the generosity of a wealthy wastrel. Unmarried, he was a favorite among the ladies, though not the more serious ones, for he was a great braggart in love, bragging less about his actual accomplishments than about what he could have, would have, done— if he could.

It was difficult to imagine that Sekula, who was steeped in looting and murder, could ever give up his trade. These, plus smuggling and other nefarious pursuits, filled his whole life. However, there was no reason to suppose that he would bring shame on his Serbian faith and Montenegrin name. It was simply that he regarded the Moslems, whom he called Turks, as naturally responsible for every evil, and he held it equally to be his inescapable duty to wreak vengeance on this alien creed and to extirpate it. He considered a traitor anyone who missed an opportunity to do likewise. He had no clear comprehension of all this, but a vague feeling, which he had inherited as a murky legacy from his forebears. Living there on the border, he strengthened and developed this feeling, taking part from childhood in border raids. The heroic hatred of his ancestors was turned by different circumstances into a criminal urge. He required no reason at all, no provocation, to carry out the murder of a Moslem or to burn down his homestead.

Once, after the war, he met a Moslem on the road from Bijelo Polje to Mojkovac. They had never seen or heard of each other before. That particular road was always dangerous, thickly wooded, and perfect for ambushes. The Moslem was happy that he was in the company of a Montenegrin. Sekula, too, felt more secure being with a Turk, just in case Turkish guerrillas should be around. The Moslem was obviously a peace-loving family man. On the way they offered one another tobacco and chatted in friendly fashion. Traveling together through the wild, the men grew close to one another. Sekula later declared that he felt no hatred, no hatred whatever for this man. The fellow would have been just like anyone else, said Sekula, if he had not been a Turk. This inability to feel hatred made him feel guilty. And yet, as he said, Turks are people, too; since they were traveling together, let us go in peace, he thought, owing nothing to one another.

It was a summer day, and the heat was overpowering. However, because the whole region was covered by a thick forest and the road skirted a little stream, it was cool and pleasant. The two travelers sat, finally, to have a bite to eat and to rest in the fresh coolness by the brook. Sekula boasted to the Moslem of what a fine pistol he had, and showed it to him. The Moslem looked at it, praised the weapon, and asked Sekula if it was loaded. Sekula replied that it was—and at that moment it occurred to him that he could kill the Turk simply by moving a finger. Still, he had made no firm resolve to do this. He pointed the pistol at the Moslem, straight between his eyes, and said, "Yes, it is loaded, and I could kill you now." Blinking before the muzzle and laughing, the Moslem begged Sekula to turn the gun away, because it could go off. Sekula realized quite clearly, in a flash, that he must kill his fellow traveler. He simply would not be able to bear the shame and the pangs of conscience if he let this Turk go now. And he fired, as though by accident, between the smiling eyes of that man.

When Sekula told about all this, he claimed that not until the very moment he had pointed the pistol in jest at the Moslem's forehead did he have any intention of killing him. And then, his finger seemed to pull by itself. Something erupted inside, something with which he was born and which he was utterly incapable of holding back.

Such were the men who gave that raid its momentum and violence. They were not at all interested, by their own admission, in avenging Boško Bošković. Sekula even hated Boško as a bully, though he respected him as a valiant hero. The murder of a former clan chieftain was an excuse for unleashing passions whose roots burrowed deep into the past, and perhaps even into the nature of the people of this land. These passions were fed by their prolonged misery and travail and burst forth when the occasion came.

Attempting to find their bearings and to conquer the new times, these men seemed only to become all the more lost, more violent, and more embittered. If the times, with their sudden convulsions—their wars and destruction and outmoded ways of life—were to blame, could not these people have prevailed? Are men doomed to become the slaves of the times in which they live, even when, after irrepressible and tireless effort, they have climbed so high as to become the masters of the times?

Old Montenegro faded away, with its men and mores, while the new order failed to bring people either peace or liberty, not even to those who hoped for these and fought for them. Failing to realize their dreams, men became bad and deformed. New men were needed with new dreams. And even fiercer battles for their unattainable achievement. This is, after all, man's fate. This fate came to pass here monstrously and mercilessly. Since it had to come to pass, could it have been otherwise?

PART THREE
Tribulation and Education

In the fall of 1924 I entered the fourth year of high school in Berane, where my elder brother had already attended normal school. There I remained five years, until the end of my secondary-school education.

Lower Nahija, as this region had been known since Turkish times, as distinguished from Upper Nahija around Andrijevica, was, with its little town of Berane, generally more developed than the mountainous sheep and cattle country around Kolašin. Yet Berane lacked Kolašin's simple and clean beauty with its strictly regular houses and streets, and its exclusively Montenegrin population, homogeneous in outlook and expression. Berane was in everything more prosperous and more diverse, riper and deeper, like all the wheat country along the green Lim. There was no mountain freshness or cleanness there, nor, for that matter, the austerity that pervaded everything in Kolašin.

The region was populated by two different Orthodox groups—the Montenegrin Vasojević tribe and the Hašani, or Serbians,* who had lived there from the very coming of

* The Vasojević clan illustrates the closeness of the Serbs and Montenegrin peoples, who are actually one, racially and ethnically. They live in an area known as Nahija, where Serbs (tribes called Hašani here), Montenegrins, and their Moslem Albanian neighbors are mixed. The Vasojevići were Serbs who settled during the sixteenth-eighteenth centuries in Montenegro and kept up their fight against the Turks from its independent ground. The Hašani were later arrivals and still called themselves Serbs. The Vasojević clan fought unremittingly against the Turks, and shortly before the war of 1861, in which Montenegro under Prince Nikola attacked Turkey, they broke through the Ottoman Turk lines to join forces with the Serbs.

the Slavs. The variegation would not have been so great had
not their neighbors been Moslems, village next to village.
These, in turn, were divided into various tribes and, what
is more important, into two languages—Albanian and Serbian. This variety was reflected in the life of the little town.

The Vasojevići were the largest Serbian tribe, and they
considered themselves even at that time the purest Serbian
blood and the best stock. Perhaps this might have been true
in earlier times, when they were not so numerous and lived
in their homeland, a desolate and isolated mountain region
along the Lijeva River, where they could preserve their
racial purity. However, as early as the eighteenth century,
they began to multiply rapidly and to spread into the valley of the Lim and around it from Plav to Bijelo Polje, so
that they became mixed with other tribes. From the first
half and especially from the middle of the nineteenth century, the Vasojevići, under Miljan Vukov, waged a struggle
not only against the feudal Turkish landlords around them,
but even against the army of the Sultan. This struggle was
so stubborn and bloody that it had no equal, not even in
this land of ceaseless slaughter and warfare. They were like
a little state unto themselves. Overrun several times, they
would then abandon their homes and flee, young and old,
into the mountains, as far as Old Montenegro, only to
return to their razed homesteads as soon as the flood of
great armies receded.

In these battles the Vasojevići were the more important,
both in numbers and martial spirit, even though there was
also a rebel movement with distinguished men among the
Hašani. The commanding positions were almost all in
Vasojević hands. This was the kernel of the idea that whoever was more active in the struggle for freedom ought to
govern later, as though the struggle were waged just for
him. There was no little tussling between the Vasojevići
and the Hašani, though it never went very far. A Vasojević
leader regarded it as an act of singular generosity if he

declared that the Hašani, too, had great men and good heroes.

Until 1912 the majority of Hašani were ruled by Turkey. Withdrawn into themselves, they were resilient and self-sacrificing whenever the need arose. They were not at all like the Serbians of Peć and Prizren, who stayed behind when the Serbian people migrated northward from Kosovo and Metohija under the Čarnojević patriarchs. They lived in towns, were few in numbers, and had been harried by Turkish terror and by the savagery of the Albanian tribes which descended like a cloudburst from the mountains to lay waste the lands of Kosovo and Metohija. These Serbians were in no position to organize uprisings, and developed a submissive and almost slavish nature. But the Hašani, like other peasants who stood their ground, as in the mountains of Lower Kolašin, had a belligerent and venturesome streak, always ready to take up arms and to make any sacrifice.

In Berane the variety was even greater than in the environs, with some gypsy and Vlach families as well as the Orthodox and Moslems. Especially on market days, the various costumes, languages, customs, and faiths wove into and yet contrasted with one another—all fated to live together, and yet hating and biting each other. Several groups and several epochs confronted and jostled one another. The town of Berane was a living picture of that stirring and brewing.

Until the second half of the nineteenth century there was nothing there but a nearby village of the same name. After 1862, when Hussein Pasha's invasion of the Vasojevići died down, Turkish military strongholds were erected on the left bank of the Lim. These garrisons, forts, and towers, amid a picturesque green countryside of small houses, fields, and gardens, offered ample proof of the battles of the Vasojevići. The establishment after 1862 of a permanent Turkish military garrison, which remained until 1912, resulted in the founding of a town a bit upstream.

The Turks had always been very careful indeed about where they erected their cities. But that was long ago, before there were such rebellions and wars. The present Turks did not care about the future town, only about their military camp, which enabled them to hold the left bank of the Lim and sink roots into the soil that the rebels had wrested from them. They established their fort on a commanding spot, both as a bridge and as a stronghold. Next to the fort was Jasikovac Hill, jutting out of the plain as though made for defense. This was no place for a town, and it sprawled, without any order, spreading over an unsuitable terrain. The Lim constantly ate away the fields beyond, and sometimes the gardens and orchards of the town itself. The town could not expand on either side; below the fort was the parade ground, Talum, and, above, the Lim stood in the way. At last it slowly sank to the nearest terrace, all crowded together, with its black wooden roofs like a flock of crows that had perched by the riverside.

On the other side of the Lim was a settlement called Haremi, which got its name because Turkish officers kept their wives there; the women were safer there in case the rebellious rayah should suddenly attack. Even the smallest town has something in common with the largest—a contempt for the suburbs. Berane looked down on those hundred or so houses in Haremi.

Despite everything, Berane grew rapidly from year to year.

The various faiths and origins placed their stamp on the town. There was a difference in the houses—the open one-storied dwellings of the Vasojevići next to the forbidding two-storied Moslem houses, while over them all towered the graceful stone mansions of the begs, with their ornate cornices and gates. As the town grew, one could detect how it became more and more Vasojević, as though the others had had their day.

The Moslems would not have had a majority in the town had not a significant portion been gypsies in the gypsy quarter, where scarcely one Montenegrin house stood. Both the Montenegrins and the Moslems of the Serbian tongue held the gypsy quarter in contempt. Many living there were not even gypsies, but a mixture of Moslem riffraff, real gypsies, and that unfortunate people the Madjups, who, as legend has it, were driven into Metohija at the beginning of the eighteenth century after a peasant revolt in Egypt. They are the most detested and most wretched people in the maze of Balkan nationalities. Many natives of the gypsy quarter had forgotten their own language and origin, and simply regarded themselves as Moslems, though not even the gypsies would accept the Madjups as their own.

There was not in all Kolašin, nor anywhere else, the poverty, dirt, and ignorance of the gypsy quarter in Berane. How did these people manage to live? How can a child possibly survive in that filth, in those rags and sores? The inhabitants of the gypsy quarter were completely stagnant and doomed to be only servants. They did not attend even the compulsory elementary school, because they found no profit in it.

The town was generally overpopulated. It teemed with students, clerks of every kind, soldiers, and officers. The bridge over the Lim was a great problem for the town as well as for the whole region. It was a wooden thing whose first half extended from the garrison to a little island in the Lim, its second half from the island to Haremi. When the Lim, like every mountain stream, swelled rapidly, the wooden bridge never did survive, and traffic was frequently cut off. Often half of the students never reached school, and the teachers did not know what to do with the rest of us. The inflow of goods was also interrupted, and there were shortages of meat, grain, and cattle.

The little town was very helter-skelter—every quarter

for itself. The only part that had some regularity was the upper and most rapidly growing part of town—the Vasojević section. At first glance the Montenegrin population seemed to be the neatest of all; they had large windows, whitewashed walls, wide streets. Actually, the neatest were the Moslems and the Serbian burghers, the old settlers. But their neatness was concealed, like everything else about them. The cobblestones of their courtyards shone, their floors were golden yellow, and their copperware gleamed. They lived a different life, in seclusion. Everything they owned was polished, soft, or warm. With them life ran its course imperceptibly, while the Montenegrins, even the more prosperous ones, lived a life of constant turmoil.

Everything else about Berane was good, and it had nature's gift, something that no one could take away from it. Like all our other towns founded by the Turks, Berane, too, stood by a river. All our mountain streams are beautifully clear and swift. The Lim is even nicer than the rest, because its cool water is fine for swimming. The environs of the town were gentle golden and brown slopes covered with fruit and grain, and beyond were the hills and mountains. Murgaš loomed black, while Kom shimmered high above like some unapproachable castle in a fairy tale. In town, cold springs bubbled, and there were gardens everywhere. The fields descended into the very courtyards. Even the streets were lined with pear trees. Berane enjoyed a mild, though mountain, climate. Such, too, were its people, especially its youths and maidens—dark and raw-boned Montenegrins together with the rather more reserved burghers; the former rowdy and temperamental, the latter mild-mannered and restrained. One day perhaps they will merge, and the people will be both gentle and vivacious.

It is out of this striving and merging, conflict and agreement, that new relations and new forms, new men and new times can grow.

Not only variety distinguished Berane from Kolašin and other Montenegrin towns; both its social and political composition were different. If there were any opponents of the unification with Serbia, they were not noticeable there. The Vasojevići, and particularly the Hašani, had always been disposed in favor of Serbia and maintained a tie with it. Karageorge, whose ancestors were said to have come to Serbia from this very region, had himself presented them with banners when they came to meet him at Sjenica. These ties continued throughout the nineteenth century.*

The memory was still fresh of the unsurpassed and incredible heroism of three hundred Vasojevići who in 1861 fought their way both day and night through Turkish territory to reach Serbia. They were the victims of an age-old and great dream—the alliance of the two Serbian states, Serbia and Montenegro, in the struggle against the Turks. The young Montenegrin ruler Prince Nikola wanted to use that march, it seems, to goad the Serbian Prince Michael into war with the Turks. By their heroic sacrifice those men demonstrated how much that dream was already a reality in the hearts of men, especially of this region. This was the reality for which these men lived and died; they left it as their heritage. Though his aim was a specific need

* See note on page 217. Karageorge (literally "Black George") was the founder of the independent Serbian state in 1804, and it was the Karageorgević family that finally prevailed as the ruling dynasty in Serbia—and in the kingdom of Yugoslavia.

of the moment, Prince Nikola's march gave rise to a great
legend. The majority of educated Vasojevići were schooled
in Serbia. The Vasojevići regarded Serbia as their land as
much as Montenegro, if not even more so. If they were for
Montenegro out of necessity and feeling, their thoughts
were of Serbia. Jealous of this loyalty to Serbia, Cetinje
regarded with suspicion even Miljan Vukov, the military
leader who finally wrested the Vasojevići from Turkey and
united them with Montenegro.

Because of all this, after World War I this region experienced almost no guerrilla action. The only bad blood that
remained was between the Moslems and the Orthodox, but
even this subsided with the political divisions that came
after the unification.

There, too, the older leaders gave way to the younger,
educated civil servants. The older leaders of this region
differed from those in Brda and Montenegro; they were
more in accord with the new times, with the unification,
and hence more reasonable. They had less of a heritage to
uphold, having joined Montenegro only recently. Their
very reasonableness may have led to the myth that the
Vasojevići were fickle, selfish, and envious, just as the
Moračani were said to be wily, with brains weighing an
oke and a half, unlike the usual brain of one oke.

Combined with this reasonableness and greater accord
with the times was the ascetic austerity and martyr-like
sobriety of the chieftains. They had just emerged from the
great and grueling rebellions in which names and reputa-
tions were made by sheer sacrifice and heroism. They differed from others in Montenegro. Civil servants rather than
warriors, they were what civil servants were supposed to
be, more civil and serving and less arbitrary, while those in
Montenegro proper were not.

Among the living Vasojević chieftains Gavro Vuković
was a truly notable figure. He was Montenegrin Minister
for Foreign Affairs, the son of the war lord Miljan Vukov-

Vešović, leader of the embattled Vasojevići throughout
the second half of the nineteenth century. Chieftain Gavro
lived at the upper edge of town. His new house had just
been finished and had cement lions at the gate. His personality brought two periods together in a remarkable way
—one of great heroes and uprisings, to which his father
belonged, and the newer one of wars for the unification, of
loud politicians and bureaucrats. He behaved himself in
such fashion that he was on good terms with the men of his
own times, those who were still alive, while at the same
time favorably disposed toward the ways and ideas of the
younger generation, an attitude that may not have sprung
spontaneously from his heart. Nevertheless, this peaceful
and retiring man was a link, thin but unbreakable, between
the most disparate men and times. Perhaps he was simply
that kind of man, or perhaps the times destined him to play
such a role in his region and clan. Perhaps both.

He was of less than medium height, already old and
feeble. A quiet and dear man, he did not regard it beneath
him to speak even with children. Despite his withdrawn
attitude, he had something all the more dignified about him,
a consummate poise. He was the master of his every gesture
and word. One could see by looking at him that the leading
families, founded several hundred years ago, were nests of
nobles.

Chieftain Gavro never went out before noon. In the late
afternoon he went for a walk, wearing a derby and black
spectacles. He walked slowly down the street, as though
feeling his way with his cane, and then to the old monastery.
There, around that establishment of the Nemanja kings,
the bloodiest battles with the Turks had taken place under
the leadership of his father. The closer Chieftain Gavro
approached death, the more ardently did he seek ties with
the past; only some great disturbance in the weather could
deter him from taking his walk. He would return at dusk.
And at night a lamp flickered behind the curtain of his

window; Chieftain Gavro was reading or else writing his memoirs. He lived his memories as he waited for death.

Chieftain Gavro was not renowned for heroism, nor particularly distinguished for his wisdom, though he was one of the very few Montenegrins of his generation with a complete education. Even so, he was the true product of his district and clan, the son of a great father. Though he faithfully served his sovereign, Prince Nikola, he favored unification with Serbia. He nevertheless did not, like many, suddenly begin to criticize and rail against his former master. He was tolerant, also, of the Moslems, and treated the Hašani and the Vasojevići in the same way. Neither a statesman nor a man of the pen, nor even a born leader, still he had a good and generous heart. His father, Chieftain Miljan, carved himself a niche with his sword; Chieftain Gavro conquered with his heart and mind, though such qualities are not as highly regarded as heroism in this land.

Born on a cold plateau beneath the peaks of Kom Mountain, the Chieftain spent his last days in the mild valley of the Lim. He had no need to descend into the plain, for he was a man of means and could live wherever he chose. Like so many others, however, he, too, finally settled in the town that had become the capital of his tribe. As an educated man of the world he was able to bring some culture into a more developed milieu, yet he was only a tiny ripple in that great stream which coursed for a century, a century and a half, from the mountains into the plains. In that struggle the plains were always held by someone alien to the mountaineers, who were hungry for bread and land, and tired of the bare, though beautiful, crags that were their home. When the old Chieftain died, it was hardly noticed. Had it not been for his famous name, one might have merely noted how, one evening, an old man with a cane failed to pass along the street to the monastery.

There were other interesting and important men. Four such men, two from the recent past and two still alive, and a woman, a mother, from a forgotten village and clan, were talked about in the whole district.

Few knew the name of one of these figures out of the past. He was known not for his name but for his deeds. The Turks killed his son. It was a bitterly cold winter, and the dead man was brought home frozen stiff. When they tried to lay him in the coffin, they discovered that they could not fold his arms. The father took a knife and pried them loose saying, "Not even a dead man should make trouble for others." What strength must have welled up in this man at that moment. When he came back home from the graveyard to an empty house, he felt the full weight of his sorrow and loneliness. He took his *gusle* and began to sing a dirge with all his soul: "The snow has fallen o'er my fields. . . ." What pain must have welled up in this man at that moment.

The second figure, whose name is known, was the renowned hero Panto Cemov. Perhaps there were other heroes who were his equal in the fleeting fury of the fray, but none could persevere in tribulation as he could. His heroism grew with suffering. No single uprising or campaign could spawn such a man, but only a rebellious conflict of several generations. One summer day the Turkish authorities caught this guerrilla leader, Panto Cemov, and drove him barefoot all the way from Berane via Peć to Asia Minor itself. There he was to be sentenced to one hundred and one years of servitude. His guard was on horseback and he on foot, his feet in fetters and his hands in chains. He endured all this as though it were a pleasantry, and everywhere along the way, though exhausted and battered, he sang from spite and gave courage to the people. Panto's inexhaustible spiritual strength showed better than anything else that such men would be victorious and that the end had come for the Turkish Empire, and even for that secluded Turkish way of life, so full of warmth and hidden passions, and also of perversion and horror.

And that woman, Mother Tola, suffered even greater tortures than Panto. The Turks whipped her only son, the monk Procopius, on the market square of Berane. She shouted to him words of encouragement, telling him to endure and not to give away his comrades, until he expired beneath the blows. Her name, too, and her awesome heroism are still remembered.

The other two, still alive, were not remembered for anything but their friendship. They were both from the same village, neighbors, and lived exactly alike. There was even a joke about them—that whenever one of them began to beat his wife, the other would beat his, likewise, as soon as he heard the cries of the other woman, just to keep things fair and square. The two men would go to town together, dressed in the same white peasant dress, girt with the same black sashes, and mounted on similar dapple grays. They were already men of years, both around fifty, but one was rather tall, and the other stubby and squat. They always strolled through the bazaar together, and at the inn they would sit next to one another. Everyone knew that whoever quarreled with one of them would have to reckon even more with the other. And they were always good for a fight, though they never picked one themselves.

Their friendship had its origin back in the days of Turkish rule, when they had been guerrillas together and got one another out of scrapes when one was wounded. Some claimed that these shared exploits merely strengthened a mutual love that had existed in early childhood. It was incomprehensible how men who were not at all related by blood could so love one another. It would have been much easier to understand had they hated one another; after all, their homesteads bordered. It would have been understandable, too, had they just been good friends. But their love was great and they made it felt in everything, even the most trifling day-to-day matters. They were more than real brothers.

Because of this kind of love, which had been a more frequent thing in earlier times, everyone held them in high regard and esteem, despite the fact that they were not particularly pleasant fellows. They were like a remnant of something long past and distant, a folk song that still walked the earth.

Yet not even their love could survive our own times. World War II separated them. Perhaps it was because their children joined conflicting camps, or because they themselves broke up. At any rate, their proverbial love perished. What bad blood there must have been between kinsmen and neighbors when not even the love of these two, a love that embraced their whole life, could find a corner of refuge to survive a hatred whose ferocity none could have suspected.

In Berane there lived a teacher named Miraš, who attracted notice for his unusual views and attitude. He belonged neither to the old nor to the new generation. He did not belong to the middle either, though in years he was between the old and the young. He belonged to himself. Nor was he neutral, since the tenacity of his convictions might be considered somewhat comparable to the stubbornness of the local guerrillas. He belonged only to his own point of view, which never found any fertile soil in that wild and tangled terrain.

Miraš lived above all events, almost beyond mankind. This small frail old man lived only, it seems, for his school. With his pointed goatee and, in the summer, his straw hat, he gave the impression of coming from some other world. He was an atheist, a tireless freethinker who mocked at religion and the church at every step. Nobody went along with him in this, though they all liked to repeat his sharp broadsides. He was a socialist, believing that wealth should be more equitably divided in the hope that men would then have less opportunity to fight over bread. In this he was even more isolated, without a single follower. Not even his

fat wife was on his side. Indeed, the whole town made fun
of their concealed ideological warfare. That constant conflict was all the more remarkable because they got along
very well in other matters and were inseparable.

Men and events passed him by—and what men and events!
But he remained wrapped in his ideas, alone and misunderstood. He was not swayed by the idea of national liberation,
and even less so by the chieftains. He sought and saw something else. Men respected him for his perseverance, though
it was senseless, and even feared him somewhat; his sharp
words cut like sabers, though none agreed with them.

Time finally burdened Miraš with eccentricity and a
petty spite, which served merely to tickle the fancy of the
small town. After all, everything was against him. He had
nothing but a bare idea and an invincible faith in what he
believed; he was similar to those martyrs who gave no
thought to temporal victory or to their own fate. He believed that his idea was the truth and that it would triumph
sooner or later, even if he never saw the day and was forgotten.

He played no important part whatever in the lively and
sharp political struggles around him. He stood above them,
cynically laughing at all factions, denying them all any
right to speak in the name of the people. As for the people,
he saw them as something brutish and steeped in ignorance.
He held it would take decades and decades to lift them out
of their raw savagery.

The dictatorship of King Alexander, established on January 6, 1929, never touched this man.* He never concealed
his opposition to it, and continued to remain what he was.

* King Alexander and the dominant Serb parliamentary party that
supported him were from the first opposed by the Croats, who sought an
autonomous and separatist position in the state. In 1928 the Croatian
leader Radić was assassinated in the national parliament by a Montenegrin delegate, the Vasojević clansman Puniša Račić. When the Croa-
tians withdrew from the parliament, Alexander declared a dictatorship in
January 1929.

He appeared hardly to notice any great change. Such a man was incomprehensible in Kolašin and in Montenegro. He seemed to come from somewhere else, though he had been born there, not Miraš the Montenegrin, but some incarnation of an alien idea which, though barren, had somehow survived. He died rejecting the last rites. That was his last act of faith. But his wife, true, also, to her own faith, gave him a Christian burial.

A kinsman of his, also a teacher, was his very opposite. He was a fiery nationalist. He vaunted his own merits, pushed himself forward as speaker on all the holidays, and threatened his opponents with a cane. His voice could be heard from the very edge of the market place as it rolled out of the tavern in defense of King and Fatherland. Miraš used to say of him that he would foam at the mouth with his Serbianism, but only before other Serbs. It was the general observation that he had not been quite so loud when the Turks and Austrians were around.

When, in the fall of 1925, King Alexander and Queen Marie passed through Berane on the occasion of reconsecrating the remains of Njegoš, certain distinguished citizens were presented to them. Miraš's kinsman had his turn and began: "Your Majesty, I am the first swallow . . ." He wished to say that first swallow which signaled the liberation and so on. But the impatient king interrupted him. "What is your profession?" Ever since then this boaster was stuck with the nickname First Swallow. Hot-blooded and pugnacious, he was known to fight over it.

First Swallow greeted the dictatorship with enthusiasm, emphasizing that it was a native son, Puniša Račić of the Vasojevići, whose revolver shots, when he killed Stjepan Radić, the Croatian leader, in the Parliament, saved the Fatherland. He became the mouthpiece of those loud tipsy village and town brawlers whose source of income was always a mystery, and yet who always had money to spend in taverns, men shunned by decent and industrious people.

This is how World War II found him, greeting the forces of occupation with bread and salt while annoying even the Chetnik war lords by his nationalistic speechmaking at rallies.*

So it was that the lives of two men from the same family and same environment took two different turns. This was rare in Montenegro. But Berane and its surroundings were less bound by the ties of clan and tribe. Men there took sides politically more easily and more fully; they acted as individuals.

* Chetniks were the followers of Draža Mihajlović, the Serb general loyal to King Peter II. His guerrilla forces resisted the Nazi invaders in World War II but were later accused of collaboration with the enemy, most particularly as a result of their opposition to the Communist Partisans led by Tito.

In Berane one could see rich peasants, as many as two or three in each village. There were prosperous peasants around Kolašin, too, yet fewer. But the rich peasants of Berane were a lordly kind, with beautiful clean houses, large gardens, and fine clean clothes, not the common variety of sheep raisers steeped in grease.

When they came to town on market day, they caught everyone's eye. They purposely set their horses to prancing and cantering into the market square. In the coffeehouses they never sat with the peasants; they joined the educated men. They, too, might have an educated man or two in the family. They rubbed shoulders with the educated and educated their children not, as with the poor, to better their lot, but to satisfy their craving for something loftier and better. This craving developed in them, to be sure, as they grew more prosperous and acquired a taste for a clean and easy life.

In Berane itself the rich and the poor were even more separated than in Kolašin. Not even there, however, were the more prosperous merchants and tradesmen very active in politics. This was largely the affair of the educated, predominantly from the more powerful clans. Wealth and business connections were too weak by comparison with the powerful clans and with officeholders. Occasionally even, some rabble rouser, let alone an energetic politician, could pull more weight than the prosperous and thrifty. Of what did the wealth of these merchants actually consist?

At best, a little house, with a store and warehouse on the first floor and an apartment of three or four rooms on the second floor; then perhaps a field, right next to the town. Any politician could have had the same from the pay of his post, without any of the risks involved in business. This, then, was the kind of beggarly wealth that these men defended in wars and revolutions, just as the prosperous peasants defended theirs, with a stubbornness and ferocity that were the greater to the degree that they had the slightest chance to develop into real capitalists.

The old merchant families were already by then in decay, regardless of differences in religion, and new families were coming to the fore—Montenegrins who had swarmed in from their villages and begun to engage in business only after the expulsion of the Turks in 1912. They were more intimately a part of the new order and in every way more adaptable to the new circumstances.

The decline of the one and rise of the other showed in their appearance. The former were, to a man, slight, pale, and mincing; the latter were strong, full-blooded, with broad gestures and step. The one lived in enclosed courtyards and gardens, languishing and pining away unseen, while the others displayed their property and boasted of their luxurious wealth. It could not be said that the newcomers were abler merchants. On the contrary, their predecessors were. They had a fund of experience accumulated through the generations, the ability to gauge the value of goods at first glance, to attain wealth by piling penny on penny and hiding them from prying eyes. They were also men of their word and dreaded to burden their souls with sin. The newcomers acted differently, more boldly, unafraid of ruin, unconcerned about honor or shame or the salvation of their souls. As a matter of fact, many of the newcomers were ruined quickly and reduced to beggary, but others cropped up overnight. The old-timers, meanwhile, slowly and silently and imperceptibly faded away.

The enriched newcomers were a magnet for the officials, and vice versa, as though they had been born together. The old local tradesmen and artisans had the same submissive attitude toward the officials as in Turkish times—authority is authority, power, no matter who wields it. The authorities, in turn, treated them with hauteur as second-class citizens. They did not dare treat the newcomers in the same way. They instinctively felt that these new tycoons were catching up with them, and would perhaps even surpass them in power someday. It was no wonder at all that the newcomers and their sons were all later to join the Chetniks. What was strange was that so many Communists, prepared to die even without victory, should have come from the older families.

The Gogas were one of the older, Vlach families. In search of business, they had moved from Prizren in Turkish times. They sensed that the new town and its business were growing rapidly. They had once been very rich, having dealt in moneylending and trade. The new blow came, it seems, with Turkey's downfall in 1912. Huge shipments of cattle, which they had supplied to the Turkish army, were never paid for. They had a large house with a garden, and lived rather isolated from other people. They were different even in appearance, soft and blonder, rather slight, and their women had strikingly pale complexions and red lips.

Even though they had suffered great losses with Turkey's collapse, it was not quite clear why those who had begun with less than they had left should have passed them by so quickly. There seems to come a time for families to decline when they have been at one pursuit too long.

The Lazarevići, on the other hand, were a *nouveau-riche* family. They started with a bakery, then bought some fields and gardens around the town. They got rich so fast that it took the whole town's breath away. They became friends with the authorities—the district chief and tax assessor—while they paid little attention to the political parties.

One of them, blond, all sinewy, was especially agile. He no longer kneaded and baked. Two workers did that. He just sat, sold, and supervised. The girls hated that blond baker, though he could have married the prettiest of them. He was not interested; he hoarded, never could get enough, and lived alone as a miser, never lifting his eyes from the cash register. During nearly one whole year I used to have breakfast at his house, and not once did he smile at me or make a joke—as though he were not a merchant at all. Nothing interested him, nothing was sacred to him, save money. But not even this love of money was pronounced. It was as though he did not know what else to do, and so he gloomily hoarded coins and looked through his little window with cold green eyes, unable to escape himself or his shop, the magic circle of tiresome work and endless amassment of wealth.

There was at the same time an unusual merchant, Milikić, who was not himself directly engaged in business, but who lent money at interest. Milikić was completely hairless and without children, a man already well along in years. He lived on the upper edge of town, in a big house half of which belonged to him. He resided alone with his wife in a room on the ground floor and lived in dread of being robbed. He was completely illiterate, but he could tell exactly whose I.O.U. was whose. He could take the whole file of notes and identify the owners without an error.

Milikić was set for the most bitter quarrel, and even a duel, if anyone joked about his hairless state. To be hairless was considered shameful among Montenegrins—a man who looked like a woman. Apparently it was for this reason that Milikić always spoke so loudly. One could hear him a block away. And he had a shrill voice, like a woman's. He would butt into conversations, like a chieftain, and even his dress was that of a chieftain, the ceremonial Montenegrin costume. It made him look more manly—grave, important, and mightier. In war he would always press forward to assert his manhood. But he had no luck; no bullet would have him.

Although people respected him because he was a man of his word and heroic, they hated him because he was a usurer. He grasped at everything he could, without knowing for whom, since he was childless. Milikić had been a peasant, but became a city slicker. He combined within himself a traditional heroism with moneygrubbing and, to his great misfortune, manliness with womanliness. All of this was inseparable in him, though it was not fused, but side by side, neck and neck, playing leapfrog.

The merchant Vučić Vujošević was also from the village. He, more than Milikić, expressed the traits of both heroism and money-making. True, he did not deal in usury, but in trade, bearing that disharmony, a war hero who traded, a Montenegrin brave and a shopkeeper. Such was his dress, too; he wore behind the counter a gold-braided tunic with a pistol in his sash, in full panoply. He knew how to put life into his business, and soon erected one of the nicer houses in the center of town. His sons were completely different from one another, as though each had inherited but a single trait of his father's. The elder was only a merchant, and the younger a hero. The elder was soft and pallid; the younger ruddy and muscular. Nothing could keep the elder from his business, not even war or revolution. The younger took up arms on the very first day of an uprising; he did not wish either to go to school or to sit in the shop. Had the war not come, he would have led a vain existence. In war, he found himself. And the war found him. He was one of the best fighters in the First Proletarian Brigade, which was not easy in a unit in which cowardice was not imaginable except as a joke. As he had cut down others with his submachine gun, so he, too, was cut down, faithful to that other, heroic, strain in his father.

Some rose while others fell. Fathers were separated from

sons. But the fathers could then hardly suspect what lay in store for their children. Nor could the children know that their paths would be different from those of their fathers. Still less could the merchants foretell their own ruin. Both fathers and children, and everyone else, traveled a common path at that time; they aspired to something better within that society.

That unity was temporary and illusory. It would seem that families, like nations, are more united when they are not compelled to unity.

My brother and I, boys from the village, were pleased that
we were not living directly in the gypsy quarter but a
certain distance away from it. Nevertheless, our house did
not differ much from those in the gypsy quarter—just a
mud hut covered with whitewash.

The house had a thatched roof, and no toilet. Behind it
was a space, no bigger than a table, enclosed by a fence,
and in it there was a box for refuse, which was from time
to time taken to a nearby field. In front of the house, in
something that might be called a garden if it were bigger,
were two or three short rows of onions and a little fruit
tree, which never bore fruit. The entrance and the garden
were surrounded by a fence of twigs held together by wire.
For water we had to go with buckets to the neighbors, who
had a well. Everything was crowded and poor, but not
without a certain warmth, which comes of work and care.

In the little house there were three rooms: a roomy one
in front, with a pounded dirt floor and a hearth; a little
room with two small windows, which was occupied by the
owner, our maternal great-aunt, Baba Marta; and another
little room, very cramped indeed, similar to the second.
Here we resided. That tiny room contained only a small
wooden bed, on which both of us slept, a rickety little
table, a bench, and a small tin stove. The room was so small
that one could just barely pass between the bed and the
wall on each side. The door, floor, and molding, were
rotten and worm-eaten. Some of the panes of the little

window were pasted over with paper. Our window looked out over the yard at a similar hut, and beyond it at a wooded round hill in the distance and the bare fields, behind which the Lim coursed unseen.

It was in that little house that I spent all five years of my schooling in Berane, first with my older brother, and then with my younger. The local authorities made haste, after World War II, to place a marble memorial plaque on this hovel; this was as senseless as when they took it down not so long ago (after my clash with the Central Committee of the Communist party and the campaign begun against me in January 1954). Apart from all this, that tiny poor hut is dear and lovely in my memory. There I spent my youth, thrilled to poetry and my first real loves, my first self-realizations and great expectations.

My brothers and I ate very simple fare. Food was brought from home on horseback—cheese, beans, and potatoes, even flour, because the bread baked from it was cheaper than that from the bakery. Baba Marta prepared meals for us. The choice was limited—beans and bacon, potatoes and smoked meat, cornmeal mush and bread with white butter. Yet all this meant we had full stomachs, unlike many other pupils. Baba Marta never took any of our food for herself; on our insistence, she would agree to take one of her grandchildren a slice of bacon or a spoonful of cottage cheese. It was not a sin, she believed, to treat her poor relatives to something, as long as we knew about it. She was a very conscientious and independent woman.

When my maternal grandfather, Gavro Radenović, fled with his family from Plav to escape a feud, his sister Marta did not go with him. Instead, she married an artisan, a potter or saddler who later moved to Berane and built this little house. Marta had had many children. Now only her oldest daughter was alive. Both of her grown sons had been killed, caught up in the nationalist movement against the Turks and Austria. Her husband, too, was killed in the war. Her younger daughter, a beautiful and unhappy girl, died of tuberculosis just after that war, having fallen ill from sorrow over her beloved brothers.

It was most strange for us, Montenegrins, that Baba Marta mourned no less for her daughter than for her sons. Montenegrins generally did not mourn much over female children. In fact, when someone wished to emphasize his grief for one of them, he usually said that he was as sad as for a man-child. Marta spoke mostly about this daughter. Hers was the last and a recent death in the house, by God's hand, after a long illness and much suffering.

Baba Marta lived alone. She even liked her loneliness after the death of all those dear souls that had departed from her. Once a week, usually on a market day, she went to town, and visited her other daughter and grandchildren on the way. One of them would drop by over the weekend. Having lost all she had, and now past seventy, she enjoyed being alone with her memories.

Marta had a war pension, and also received money from us, for rent and her cooking. This is all she had to live on, and she could have gotten along with less. Yet she always had coffee on hand and was neatly and well dressed. She had once lived better, but knew that nothing comes free, that one must spend one's life in toil. Baba Marta was rather tall and still slender in the waist. Her green-blue eyes had not yet grown dim, despite the many tears they had shed. She dressed like the Moslem women, in baggy trousers and cloak. From 1912 onward, however, she did not wear a veil. Like other Orthodox women of Turkish towns, Prizren and Peć, she had Moslem ways, but she was more liberal in outlook.

Her great and long-standing passions were tobacco and coffee. She seemed to live on only these two and some greens. With remarkable dexterity and ease, in darkness as well as in light, she would roll a cigarette on her knee and then moisten the edges of the paper with her lips to keep them pasted together. She always drank her coffee bitter, holding half a cube of sugar on her tongue. The tobacco she smoked was good, from Scutari if possible, yellow and very finely cut. She enjoyed both, savoring them slowly, after the frequent manner of older men in Moslem towns. She experienced the savor with all her soul, with a deep inner tranquillity and quiet contentment.

Not only had Baba Marta long ago become unaccustomed to peasant ways, but she looked down with contempt on all villagers, even the greatest Vasojević chieftains and rich farmers, because they were, as she said, peasants and uncouth, clumsy in their movements, rude in their speech, devoid of any fine sensibility for smells or colors or the taste of food and drink. Except for very special guests, none could enter her room save in his stocking feet. She could not stand our peasant sandals and mud-covered stockings, our homespun clothes, our sniffling and spitting.

She did not own a sharp tongue; she knew how to express with a gesture, a glance, or some other sign what she thought and felt. This did not mean that she did not like or cultivate gentle speech. She simply regarded words as beautiful things to be used rarely and sparingly, to be enjoyed. Thus, on rare occasion she would utter some lovely expression or a verse of some poem. Having lived long in town and in the manner of the town, having mixed for decades with confirmed city women, both Moslem and Orthodox, Baba Marta was a very rare example of a woman from an old-fashioned mercantile and feudal Turkish town in which it was religion and vague nationalist aspirations that divided men most, while everything else was the same and shared by all. The Orthodox townspeople were closer in way of life, and often in outlook, to the Moslem merchant and artisan than they were to the Orthodox newcomer from the village. So it was with Baba Marta. She adhered firmly to her faith, and clung to her ways, like those of her fellow citizens of other faiths.

Loves and passions, sickly and intense to the degree of complete intoxication and even death, were frequent in such an environment. Impetuous begs who got themselves and their horses drunk found the streets and yards too cramped. The sons of merchants were sad and morose for lack of knowing what to do with their time and strength. Beautiful plump maidens and young widows caused serenades to be sung to the lute, knives to stab breasts, and whole regions to go mad. All of these things were, as is known, a part of life in these little towns in Turkish times. There appeared, also, other, concealed, amours and infidelities, over which one pined and wasted away in yearning or despair within walled courtyards. As is usual in any decadent society, the men of that time lived for great and refined passions.

It was not quite clear just how Baba Marta's life stood in this regard. She hid this part of her life, like the other, in accord with the mores of the milieu in which she lived. The older men recalled that she had been a very beautiful woman. Judging by her attitude, which was not at all strict, toward amours, not only of young men and maidens, but even of married women, one could conclude that her life had not gone by without great yearnings and gripping passions. Apparently during her youth and maturity the times were less strict respecting love. Public scandal was not tolerated, yet secret deep loves were an aim in life, and easygoing flirtations were a form of entertainment. To cast a glance at a young man, to wait night after night for a girl to appear at a window just to see her silhouette on the blind, or to listen for a whisper through the gate, was quite permissible and needed to be hidden only from one's father and mother. Some of that still remained. In that little town there churned a lively but concealed love life among the young people, unlike purely Montenegrin towns, which were stricter and soberer in this respect. The play of love was something inherited here, something that developed quietly and came increasingly into the open.

Baba Marta never condemned a girl for falling in love with a man. She did not consider it a great tragedy if a fellow left a girl—she could always get over it—but nothing of the sort should happen after she was engaged. Baba Marta looked upon marriage brokers as a quite natural institution, without which there would be no marriages. How would young people ever get married if someone did not lend a hand? But to gossip to anyone about one's ventures as a marriage broker was very poor form and shameful.

If any married woman openly overstepped the bounds, Baba Marta would condemn her. But love on the side, hidden and hushed, enjoyed her sympathy. She regarded the married state as something wise and profitable, a necessary part of life, for people must have children, a house, and a helpmate in this world. However, love, too, was something even more inevitable, especially for a girl before marriage. This was the sacred right of youth, moreover, of every human being. Without it our greatest and most beautiful wish in life would never be realized. A life without love was only a yoke to be borne.

Life in a Turkish town, with its coquetries and love, which grilled windows and courtyard gates could not hinder, had an all-pervasive pulsation, which one could not help but feel. In such an easygoing life, yet one so filled with emotions, human life was not so short; it had an inner order that only wars, uprisings, and great disasters and plagues could disturb.

None of this meant that Baba Marta liked the Turks and their rule. Far from it. She simply could not tolerate the disruption of this tightly woven way of life, which generated from within itself its own emotions and passions, and which could be suppressed and broken only by violence and catastrophe. Though she regarded violence and catastrophe as something unnatural, she looked upon them without fear or horror, as something inevitable, without which

life could not be. Side by side with that easygoing life in
Turkish towns, it seems, there were also horrors of every
kind. Usually they burst in from the outside, perpetrated
by Turkish soldiers and officers or berserk feudal lords.
Sometimes, though, they sprang up from within, and then
caught everyone in that languid environment by surprise.
Tranquillity and warmth existed side by side with violence
and perversity. Both were but two, inevitable, sides of the
same way of life.

Much of what seemed horrible beyond comprehension
to the new generations was for Baba Marta, as one could
gather from her tales, if not understandable, then at least
common, for it was a part of everyday living, which, however deformed, existed beside tranquillity.

A peasant woman from Vinicka was passing by the
stone watchtower overlooking town when two men started
out after her. She must have been an energetic woman, or
perhaps shame gave her strength and determination. She
began to run. They caught up with her on the ledge that
hung directly over the town. Nearly half of the town
looked on as the two men raped her, up there within the
sight of all.

This was such a horrifying scene that I could never
look at that spot without recalling that unhappy peasant
woman from Vinicka, with her wide skirts and many-
hued stockings. What Baba Marta and, it seems, the other
townswomen of that time found so horrible about this
was that it should have occurred within sight of the
whole town, without regard for shame and the decent
feelings of the citizens. The deed itself shocked them less;
after all, it was not so rare and therefore not such a terrible
phenomenon in those times.

There lived an old man in town, a town constable, who
had been in the service even in Turkish times. He was a
Serb, though apparently not from those parts. Rumor had
it that he had once been one of the handsomest men of

his time, as fair as a maiden, they used to say. Traces of that beauty were still visible in his big, though beclouded, eyes, in his ruddy face, in the blackness of his mustache and hair, even though he had already grown slightly gray, weather-beaten, and bent. That man became mortally ill, and in a raving fever he attacked his own daughter. The girl barely broke away, her clothes torn and she all bruised; she saved herself by fleeing into the neighborhood.

It was then that Baba Marta told—as a great secret, though one that did not shock her much—that this same man had been the lover of a Turkish district official, that is, he had served the Turk instead of a woman when the official came on an inspection tour of his district and could not bring his harem with him. In those days things of this sort were done almost publicly and were not regarded, as they are today, as unnatural and shameful. Just as men were impaled on stakes, beheaded, or had their homes burned down, so this, too, was another regular feature of Turkish rule and of the decadent Turkish overlordship.

What had happened inside the soul of this man in the course of some thirty years since he had stopped being a paramour? He married, begot children, and lived a peaceful and retiring life. Then suddenly something depraved erupted within him. Had his perverted youth reasserted itself? Or are these human passions without bottom and inscrutable?

In the time of the Turk, life was both sweet and cruel among the prosperous families in the towns. There were flaming loves and smoldering passions, but also monstrous horrors perpetrated against men, women, and children, and especially against the rebellious Christian rayah. There are two sides to life. In Turkish times they were perhaps more obvious, and inseparable in the life of this part of the country. And perhaps in the life of mankind in general.

The high schools in Berane and in Kolašin differed less than did the two towns and their inhabitants, though they bore the stamp of their surroundings.

I arrived in the Kolašin high school at a time when the postwar situation had already begun to settle.

Montenegrin children generally began to attend school later than the rest, and, because the war had retarded many, in 1919 there had actually been in the lower grades grown young men who left their guns and knives in the neighboring coffeehouses before entering the school grounds. There were among them even those who had fought as guerrillas against the Austrians. In my time, however, it was only the rare student in the upper grades who secretly carried a revolver, just in case.

Perhaps the situation in this respect was somewhat better in Berane. There had been fewer guerrillas in that region and hence fewer weapons among the pupils. Still, there must have been many more grown students, considering that there was a high school of eight grades and a normal school, whereas Kolašin had only a high school with six grades. In Berane the grown-up pupils finished school with the speed of a bullet—two or three grades in one year.

Kolašin had a solid building, constructed before the war, and there was not too great a number of pupils. Thus, the instruction could be carried out with greater care and orderliness. It was different in Berane. Had not these dis-

tricts, especially the Vasojevići, had so many influential
men on all sides, they never could have had both a full
high school and a normal school. There were not the
slightest prerequisites there for this, except the desire of
certain leaders to develop their region and the irrepressible
desire of the peasants to educate their children, since they
did not know what else to do with them. There were
enough students, but everything else was lacking—professors, buildings, and equipment.

Both the high school and the normal school operated in
a single long building, hastily erected, and in two big old
buildings, which had served earlier as Turkish *mektebs* or
parochial schools. The building, a low brick structure, was
similar to a barrack. A corridor extended along its entire
length, and from it one entered each schoolroom, one after
another. This barn was covered by a shingled roof that always leaked. The plank floor was rotten. The schoolrooms
themselves were light and airy. The other two buildings
were higher, built over cellars, each with four small rooms
separated by a corridor. These buildings were quite serv-
iceable and housed the normal school until it moved later
into a newly erected elementary-school building.

Berane had a shortage of public buildings. Though construction work went on, it was too little and too slow.
There was quite a scandal over the Temperance Home,
which was fifteen years in the building and not finished
until after this last war. The post office, police station,
hospital, and hotels were all built anew. Only the district
administration was placed in a fairly decent building, which
was once the headquarters of the Turkish district chief.
The seats of authority and the jails were always solidly
built, for people knew they had to exist. On the upper
story were the officers, and on the ground floor was the
jail. The jail became too small after the liberation, and
another stone house was taken for this purpose. After this
most recent war of liberation, this jail, too, has become too

small. They have probably erected a new one, solider and larger. In Berane there was an even greater shortage of instructors and equipment than of buildings. Some subjects were not even taught, or very irregularly.

The relations of the instructors with the pupils were brutal. Slapping was very frequent. The Kolašin high school, though better organized, was worse in this respect than Berane. Maybe the greater primitiveness of the region required this. On the other hand, none of the parents held it against an instructor for beating the children. On the contrary, they encouraged it. The instructors would have found it difficult to act otherwise under the circumstances. What else could they do with their wild pupils? There was constant internecine warfare among the pupils, rudeness toward the teachers, bedlam and pandemonium, vandalism in the school yard, destruction of windows and furniture inside the school. One could either expel the students or beat them. Actually, the high-school instructors merely applied the same methods that had been used before, and so generously, by the parents and elementary-school teachers. Such methods were quite at home in the village. The pupils were born under the rod. The children were beaten, and beat one another. Fathers beat mothers. The authorities beat the peasants, and the peasants beat one another. The rod seemed to be not only more natural in such an environment, but more profitable, to both the pupil himself and his parents, than expulsion from school.

Discipline in the school depended on the principal in far greater measure than one would suppose. The Berane high school had only an acting principal. Kolašin had from the very beginning a principal named Kesler, a very fine man, cultured and more conscientious than severe. One can imagine what tortures such a man suffered in that savage and belligerent community. Immediately after the war he had to guard against armed pupils. Later he had to fear their parents, who were no better and even worse. He was afraid, and with reason, that he might be murdered. As a stranger, he could not even hope that fear of blood revenge would deter his murderers.

Kesler was a Dalmatian, who had been moved by patriotic feelings to serve in Montenegro, to uplift culturally a backward region. His situation was, therefore, incomparably harder and more complicated than that of the Montenegrin instructors. First, he found it more difficult to adapt himself, and second, this exclusive and ingrown community treated him as a stranger. It was rumored that he was easily frightened, so many took advantage of his good nature. A pupil's parent would burst in on Kesler shouting and threatening; a pupil would hurl at him the insult that he was from a bloody house and that all of his kin were murderers. Kesler always sought to iron out misunderstandings, except when it concerned a pupil's grades. There he was adamant.

Unlike the other instructors, he never even tweaked a pupil's ear. As in everything else, so in this, too, he adhered firmly to the principles he had adopted earlier. Fiery and a bit hasty in his speech, he never forgot himself to the point of overstepping the bounds of his pedagogical credo.

Nevertheless, Kesler got along and soon had the school in good order by dint of quiet perseverance and conscientiousness. A physicist by specialty, he was forced to teach other subjects as well. His chief task, obviously, was to put the school in order and to exert a cultural influence. Even his sons served in this respect. They were the first to bring basketball and soccer to Kolašin. Kesler realized that such work in a primitive community was only the beginning and a sacrifice for something more beautiful and better, something to come in some distant future.

Kesler was a small, thin, and very lively swarthy man with a high and bony forehead and trimmed black whiskers. He could never sit still. He made most of the equipment for physics and geometry with his own hands. He planted trees in the courtyard, and even around town. He could not rest for a minute without work and creativity. Up in the garret of the high-school building where he lived, a kerosene lamp would shine till late at night. He was working at night for the school and for a region that was not his.

Kesler was replaced in 1923 by Mirko Medenica, a native of the Kolašin district. He, too, was worthy of his calling. Medenica was a very handsome, rather small man, with black hair, whiskers, and beard, a white skin, and a pair of flashing eyes behind his pince-nez. He was not over-severe, though unyielding and particular over even the smallest trifle. Medenica had been educated in Russia, whence he had fled from the Bolsheviks. His wife, Varvara, called Varya, was a Russian, also a small person, and very pretty. He and his wife always spoke Russian. People thought they did so because it was more elegant. Mirko had manners that the town believed to be Russian. He kissed the hands of the more distinguished ladies, always let them pass first, carried Varvara's cape, and permitted her to go around town with Russian *émigrés*. Though he spoke our language well, sometimes a Russian expression would slip in, especially that Russian *no*, or "but," which he used with every other word. Medenica was more violent in everything than Kesler. He could not refrain from pulling a pupil by the hair or giving him a sound cuffing. Fulfilling his duties with Kesler's conscientiousness, he was at the same time a son of his native land, quick-tempered and sharp, though he had something in him also of Russia, where military discipline in the schools was elevated to a pedagogical principle.

In Berane the acting principal was generally Professor Dragiša Boričić. If the expression "picturesque" could be applied to a person, then it would certainly fit him exceptionally well. He knew a bit about everything and, more important, he was able to discuss everything with great eloquence and persuasion, with enthusiasm and passion, as though that was the very thing he knew best and to which he had devoted his entire life. He spoke fairly good German, French, and Turkish, and dabbled in history, and even in physics. Once, a certain hypnotist, a trained doctor otherwise, came to the school, but it was Dragiša and not the doctor who gave a lecture on the secrets of that art.

His field was Serbian language and literature, but involved as he was in his duties as acting director and a jack-of-all-trades anyway, he rarely taught his subject. More often, he substituted for colleagues in other subjects. He had been a student of the famous literary historian Skerlić, and one of the most talented, at that, but carried over nothing from his teacher except a romantic nationalist fervor, without any sense of moderation or measure. That fervor possessed him still, ten years after the war, which was a bit funny, yet all that remained were resonant and hollow words.

Boričić had taken an active part in the effort against Turkey and Austria. He had even served as a guerrilla in Macedonia. All of this seemed to be, in his case, rather superficial, done in the sweep of enthusiasm. It was in this easy and romantic way that he enjoyed the national traditions, folk songs and tales, and the poetry of Dučić and Rakić. He referred to his great teacher Skerlić with a rather too obvious intimacy. "Poor Skerlić," he said of his recently deceased teacher. In those two words there was real sorrow, as when men talk of their lost youth. Maybe that is how it really was; instead of the nationalist dreams of youth ensued postwar selfishness and greed and the crudities of small-town life. The old aspirations and dreams had been unreal and false, but one could not, dared not, admit this. One had to live for something, after all.

He may have been over forty, maybe less. That is how he looked and, moreover, how he acted. He was straight, tall, and strong, ruddy, with a prominent nose and chiseled features, big green eyes, even though he was dark, and evenly gray hair, which seemed from a distance to have a greenish tinge. It was his hair that was his most remarkable feature, and he knew it. It was soft and wavy, cut in a circle, and combed up so that his massive head looked all the bigger. He often wore a butterfly cravat, the big kind actors wear; it went well with his hair and his whole unrestrained and exalted pose.

Apparently, the most important thing for him, as for most people, was to present to the world an imagined self, which had no connection at all with the real personality beneath the mask.

Boričić's greatest concern was to look like an artist, and he acted and dressed accordingly. He acted, recited, played the violin and flute, dabbled in archaeological digging, wrote memoirs, poems, and stories, gathered anecdotes, made speeches, and even sketched a bit. Always in a flurry and enthusiastic about everything in life and in school, he never took up anything for a long time or thoroughly, but grasped at everything along the way, unable to dig deep into any one thing and to master it. He scattered his talents on all sides, unsparingly, and this is why they turned out to be slugs instead of ducats. Had this man of many talents seriously taken up any one thing with an inner conviction and perseverance, he might have become something. As it was, he was something that flashes and passes quickly away, a lighted torch of straw.

He was born somewhere near Berane. One could sense that there was a vital bond between him and the local traditions and local people, especially those from the village. Tall and conspicuous as he was, he would greet peasant men and women on market days with ostentatious cordiality and call out to them across the street. Unlike most intellectuals of that region, who dreamed of being members of Parliament, he did this not out of political calculation, but with that same romantic and popular enthusiasm that was, however unrealistic and contrived, like

his personality, the essence of his life. He was, after his
own fashion, a social leader, but not a political or religious
one. He regarded himself as the poet laureate of the community. In fact, he served as such on every occasion that
could not be imagined without poetry—with eulogies and
odes of praise on festivals and at receptions of cultural
and political dignitaries.

Once, a training plane was forced down on the field outside town. A huge mob collected there out of curiosity.
Boričić availed himself of this convenient opportunity to
excoriate our ancient foes and to exalt the heroic defenders
of our blue skies—in this case two indifferent and quite
simple sergeants. On every such occasion, out of his torrent
of empty words there would leap up some beautiful phrase,
which would be long remembered. To those two pilots he
cried out that they did not even suspect what concern and
love accompanied them as they plied over these cruel
mountains. He was witty at his own wedding, too. As the
crowns were being placed on his head and his bride's, he
remarked that they looked like Tsar Lazar and Tsaritsa
Militsa. This was a bitter jest at his lost bachelorhood.

It was rumored that he was quite a ladies' man, and not
a very particular one, at that. Some widows, who were
neither young nor pretty, boasted of his nocturnal visits.
The older students, who already played cards and chased
after women, would run into him late at night as he was
returning from his nightly prowls. Strong and energetic,
he would run after them over fields and ditches, but because it was dark and the town was poorly lighted, with
only a kerosene lantern here and there, neither could
clearly recognize the other. As soon as they caught sight
of a big fellow who paused to get a look at them, they
knew that it could only be he. As soon as they began to
run, he knew that they must be students, and so would
rush after them. Women of good reputation avoided being
seen with him in public. However, they rather liked to

have it rumored that they had repulsed Dragiša, for he was the only local celebrity—a poet, lover, and national hero. While the more mature and serious ladies adopted this attitude toward him, the young girls peeked at him behind corners and window curtains with curiosity and shame and timidly fled from him as from a sorcerer. He was a dangerous old bachelor, sly and irresistible, who could seduce one with his eyes and intoxicate with his speech.

His actions and methods as a teacher were also unusual, and, above all, unexpected. He never punished anyone by expulsion from school, but he used lesser punishments all too frequently. He resorted to these swiftly, of course, and always in a different way. Violations that would ordinarily bring the miscreant a good beating sometimes provoked only a scolding and curses—yes, curses—and a twisted ear. Boričić seemed to delight in surprising the pupils in everything. In fact, in his punishments, as in everything else, he was an improviser. He would hide in the bushes behind the toilet, which was by the stream behind the school, to catch pupils smoking. He would give them a good going over and take away their tobacco pouches. He had the habit of eavesdropping at the door of the most unruly classes, which the teachers, usually women or Russian *émigrés*, could not control, and then he would burst in when the bedlam was at its zenith. On the street, if he could not catch an erring student, for there were those who would not halt, he would throw stones after him. He was never rough toward the girl pupils; they were, after all, the fairer sex, yet he never carried on any open flirtation with them.

The instructors did not like Dragiša Boričić. Those possessed by a desire to become authors, or to distinguish themselves as orators at funerals and festivals, never made it because of him. Those who were poor in their learning sensed his superiority and feared him. The serious and educated instructors considered such a man inconducive to the maintenance of order in the school. And everyone was affronted by his brilliant and widespread fame.

He was very much liked by the pupils because of the vitality of his temperament and the romanticism of his whole personality. He loved the *beau geste* and the *mot juste*. Every pupil could feel somehow that he would find unexpected understanding in this man. And so it really was. In school he was incapable of establishing order; he had not the talent or urge for it. Under his direction one could feel that the pupils exhibited a certain exhilaration. Just as his joys and enthusiasms were shallow, so it was with the maintenance of order and discipline, which bring gloom and sadness. He was not a man of sorrows, even though he shed poetic tears over everything from the Battle of Kosovo to the widow's orphans.

Boričić was essentially a good and noble man but an amateur, and deeply unhappy. He had realized nothing of what he had loved and desired. Why did he take frequent, long, and sudden walks in the fields? Why did he stand so long before the Lim, staring at the capricious play of its eddies? Why could he not stick to any one thing? His being was lonely and oversensitive for that time and that place.

Still, he accomplished much for that region and even more for the school itself. Though he did not establish order, he did give the school impetus under conditions of poverty and turmoil. By his efforts he enabled the region to give rise to several hundred intellectuals. Perhaps they were not the best-trained men, but yet compared with the peasant primitiveness from which they had sprung, they were something. They were, for that region, the beginning of a new era after the long Turkish rule, at a time when educated men were rarer than church bells in a Moslem land.

Boričić lost his life somewhere in the Sandžak. The local Partisans shot him during the last war. I hardly think that he deserved to die. Sadder than death was the fact that his end bore the stigma of public shame. With him, despite everything, was shot a beautiful memory. There was something inevitable in that death. Disappointed in the achievements of the national struggle for unification, he nevertheless remained faithful to them, at least in words, while he succumbed more and more to the grim realities of everyday life. Such is life, as the saying goes; what can one do? One had to take as much as possible from life to compensate for the irreplaceable, for what was lost in the wars and in struggle. Such was the destiny that overtook the generation to which he belonged. Taking as much as they could from life, they lost their own lives. They fought, but something else resulted which they had not expected. They died, in fact, along with their dreams and hopes, each in his own way.

Society has no way out of disappointment but the death of whole generations and whole classes, just as of individuals, no matter how much vigor and lust for life they feel.

In the Kolašin high school two instructors were outstanding—Šćepanović and Radović. Both were from the same
region.

Novica Šćepanović taught geography, as interestingly
as if he were telling fairy tales. After all, the world is en-
chanting, in every nook and corner, as long as one is able
to describe it so. And he could. But he was very strict,
even more in maintaining discipline than in grading. He
never beat pupils, though in some exceptional cases he
would pull them by the hair. His severity asserted itself in
another way; he was a merciless teaser. If a pupil yawned
without covering his mouth—and this was quite common
for peasant children—Šćepanović would start to scold him
and then imitate the unhappy wretch with broad gestures
and contorted face. This was more terrible and degrading
than any beating. Slob, dumbbell, lummox, dunce—these
were his favorite expressions. Far more intolerable, however, were his insulting voice and manner, and especially
his mimicry and relentless repetition of one and the same
grimace or epithet, more usually both together.

Šćepanović was somewhat of a morose man, in conflict
with a primitive and rude environment. He always exuded
some sort of bitterness. Above and already outside of that
milieu, he nevertheless knew it all too well not to be re-
volted by its crudities. Not only did he not conceal his
bitterness, he paraded it. Thin, frail, but tough and all
nerve, an excellent instructor and pedagogue, often misunderstood, he took out his bitterness mostly on the pupils.

He was the brother of Stana, with whom I had lived. Unlike her and her sister, Stanija, from Podbišće, though he looked like them, he was industrious and persevering, in a dogged sort of way. Education had turned him into a joyless skeptic toward life and people, and especially toward that particular community, though it could not deprive him of his native diligence and intelligence. His sisters were peasant women, and had no prospect in life. They lived in toil, taking care of younger lives. He, on the other hand, was driven by schooling, reading, and an intellectual life to contemplate and to draw conclusions about the world and life, but in an all too sophisticated and, moreover, caustic way, which pervaded all of him.

The pupils feared Šćepanović more than Gligorije—Glišo Radović—a phlegmatic and indifferent man who addressed his pupils all as "my pigeon," whether he was angry or not, and let fly with a few good slaps without many words. Glišo was a sickly man; he was never seriously ill, but it was rumored that he had tuberculosis, and the pupils had a frantic dread of coming too close to him. He always spit in his handkerchief. Red splotches on his face, the fatigue in his eyes, and the wrinkles on a still young forehead betrayed some sickness deep within.

He neither liked nor hated any of the pupils. He was that kind of man. He did not ask for sympathy, nor did he give any. He had no contact at all with the pupils as people. But he made everyone respect him, and thus fear him. He made quite a point of educating as many children as possible. He took his vocation as an ordinary everyday job and not as a lofty aim or something above him. Had he lived in earlier times he would have been just as great a hero, and in a monastery he would have made just as good a monk. In any event, he would have become whatever he started out to be, and without turning back for the rest of his life, as though there were not and could not be anything better.

He taught history, in a voice as monotonous as reading in church. He never finished his lecture by the end of the hour, but got lost on the way in some detail that was of special interest to him. Then he would ring his bell and tell us what to read.

Gliŝo was just as relentlessly and dispassionately politically partisan. Every market day he would sit in the coffee-houses with the peasants, drink plum brandy with them, and expound his views. Unlike him, Sćepanović took a very reserved attitude toward politics, like a highborn gentleman who feared that contact with peasants might get him dirty or flea-bitten, and as though this were his only worry. Both, each in his own way, contributed to the school and to the community: one by struggling with them, the other by accommodating himself. Apparently both methods were equally good, as were both men. Like everything else in life, a school consists of not only one, but many sides.

In the Berane high school, too, there were conscientious and good instructors. Among them the most distinguished at the time I was there were Mijović, Ivović, and Žečević.

Ljubomir Mijović was a professor of mathematics and physics. His conscientiousness and exactness were so consistent and complete that they would have been an intolerable torture and calamity for any other man. For him they were a way of life.

Just as mathematics is not subject to whim, so, he believed, man must not be either. And truly, if a man can himself completely turn into a science, then that was the case with Mijović and mathematics. He even computed grades in a special, mathematical, way, not only with pluses and minuses, but by introducing decimals to the hundredth. Thus in the course of a semester he would record grades such as 3.45, 2.70, 4.40, and the like. Mijović even entered these grades in his notebook according to a special code, so that it was impossible for the pupil to learn what he had received. If the total, after all the addition and division, turned out to be less than .50, then the final grade for the semester would be the nearest whole figure. For example, 3.45 gave a grade of 3; on the other hand, 3.55 meant a grade of 4. The fraction that he thus took from a pupil or gave to him had to be compensated for the next semester.

Such a system of grading was as much the consequence of his philosophy of life as it was of his professorial pedantry. The world and its laws were no more than a set of mathematical quantities and relations. By his imperfection man merely brought disorder and chaos into that world. However, it was man's distinctive characteristic to strive for perfection, that is, for order and harmony, and therefore he constantly seeks to raise himself to the level of mathematical laws. The human race was doomed to an endless disturbing struggle for mathematical perfection and a yearning for it. This was man's whole misfortune and fortune.

No power or circumstance could force this man to be partial in grading a pupil, not even an affront by any one of them. Political, family, or other considerations did not have the slightest influence on him, and toward his own relatives and fellow villagers he was even severer and more aloof than toward the rest. Partiality in grading was for him tantamount to the violation of some eternal moral law. Knowledge, measured in milligrams, was his only yardstick, and in this respect he was an incomparable exception. He made no difference in importance between written and oral work. With him it was impossible to cheat during a written examination or to supply whispered answers to a reciting pupil. He was all eyes and ears. His ears were large and always seemed perked up, as though seeking to trap sounds nobody else could hear, while his eyes were set very far apart, so that it seemed he could see both sides of the room at the same time.

It was a real miracle how quickly we pupils, unruly and

willful as we were, realized that there was to be no nonsense
with him; during all of his lecture hours we were as quiet
as mice.

Mijović was a thin, wispish man, almost bald, and with a
head far too big for that frail body. He looked like a tad-
pole. He was nervous and touchy, and especially quick to
suspect even the slightest trace of derision of his ungainly
appearance. Not even the most cleverly camouflaged allu-
sion in this direction could escape his attention. He would
then be seized by bitterness and anger, which he would
get under control only afterward, at least in time to prevent it influencing his grading of the pupil. The pupils did
not like him, except for two or three in the class who liked
mathematics, for there are such people. They were in love
with him, his conscientiousness, his preciseness, and undoubted command of his subject.

That is how Mijović was—orderly and precise—in his
private life as well. He lived modestly and withdrawn into
his family. Sparing, stern, and unapproachable, he obviously regarded the teaching profession as his only aim in
life, and mathematics as the mightiest weapon of that calling to bring order to the undisciplined and chaotic human
spirit, particularly in this wild and unruly land.

He never aired his political views and probably belonged
to no party at all before the royal dictatorship of 1929. At
the beginning of the dictatorship he remained silent. He
was the only one among the instructors who gave no sign
of any great change having taken place. He simply continued to carry out his teaching duties conscientiously and
patiently, like a priest celebrating Mass come what may.

Later he became quite hard on the Communist students
—not in his grades, but in ridding the school of them. He
was transferred to Serbia, where he supported fascist or-
ganizations during the war.

What happened to this man who loved only his subject
and his family? Perhaps his mathematically orderly cosmos

fell apart, and he could never comprehend or accept it. Actually, he never understood that in society, even more than in nature, there exist not only varying dimensions, but varying measures for the same quantities, different systems of coefficients, in which the same masses and degrees of energy receive different and even opposite values. He believed that there existed for mankind but one, already established, system—his—and that everything else was merely a chaotic disturbance of that order and those already established values.

After the war he was, of course, dismissed, and unable to earn a living to feed all his many children. He came to seek my help in 1946. I could not understand at the time why an excellent teacher was not allowed to teach his subject further, even if he had some political sins against him. He was reinstated. But his enemies were unrelenting, even though they could not deny his qualifications as a teacher. Again they raised his case, again discussions, and he was transferred to another town. He already had some grown children at that time, who became members of the Communist Youth. He did not use this at all to keep his job, nor did he boast of it, but simply did nothing to hinder them. These were new times—let them go where they pleased.

His order had been destroyed. He was completely crushed, to such a degree that he even tried to convince others that he was not against Communism. He had been thrown out on the street, hungry and in rags, with his wife and children. His case prompted much thought. What was to be done with men who are conscientious and qualified in their specialty, but who are ideologically at odds with the new state of affairs? He was a part of that great and general problem. The opposition to its reasonable solution came to have the force of a prejudice that none could control. The new Communist bureaucratic class, though in its ascendancy, had neither grace nor understanding for anything except its own interests. It, too, had its own order, even if not a mathematical one, but one which was still narrower and more exclusive.

Dušan Ivović was, with a few short interruptions, my home-room teacher until my fourth year of high school. This proved to be very useful; he supervised the same class over a long time, helped us, and knew the good and weak side of every individual. His field was history, but he taught other subjects as well. He was an excellent teacher, conscientious, and, above all, a helpful and gentle man. It was a pleasure to watch little bubbles appear at the corners of his lips as he lectured; so intent was he in his exposition that he never noticed. Of a soft womanly nature, he was like that in physical appearance as well; he had a tender skin, a certain litheness, a pale high forehead framed in soft hair. He was very resolute in defense of his pupils and class whenever any instructor made an unfair charge. The pupils knew this, for nothing that happened in the school could ever be hidden from them, though he himself never let them know. Suddenly we would notice that the injustice had been nullified.

Yet this man, too, later took energetic action against Communist students. Until the dictatorship, he was democratically oriented. He adopted a tolerant attitude toward the Communists and was seen in their company, even at the beginning of the dictatorship. But later he changed completely. Perhaps what happened to him happened to many others as well. He swallowed some particular vile pill of the dictatorship, swallowed his pride with it, and then he began to justify his own betrayal to himself by becoming increasingly zealous and bitter. Perhaps the tide of revolution, which could be felt even earlier, threatened his accustomed way of life, his notions and dreams. Or this was a good excuse for him to turn reactionary. His gentle nature did not keep him from being severe in the increasingly relentless struggle with his political enemies. These were separate, though connected, traits in a single personality—his own.

Zečević was a good teacher. His subject was the Serbian language and literature. He lectured with passion, entering into every detail with heart and soul. He lacked any literary talent, but knew how to point out a beautiful passage and a good book. Modern literature did not attract his attention; he clung to the old, what he had learned at the university. He poked fun at our local literary lions, Boričić included, though never in class, only around town. "Woe to the village where the chickens sing," he would say. To make fun of others and to tell tales—supposedly just between him and you, eye to eye—was his incurable passion. Yet he did not indulge in it out of any spite or gain. This was simply his spiritual nourishment.

He was partial in handing out grades to pupils, but never to their loss. That is how he was in politics, too. Politics, he believed, by its very nature permitted all means, as long as they were not too crude. To blacken an opponent and to catch him thus by surprise was his favorite method. It was not profit that motivated him, but strictly a passionate inner urge.

To the older boys—supposedly in private—he liked to talk about girls, and very frankly, arousing their desire by his raw descriptions of loose hair dangling on bare shoulders, naked thighs and buttocks, protruding breasts, petting. He went so far in this that he even spoke of similar things with the girls. I saw one of them once, behind the stage curtain while a play was being rehearsed for St. Sava's Day, her face aflame from his whispering. He would even tell a certain girl how he knew that a certain boy was in love with her. He would then say the same to the boy. He would interfere in suppressed and naïve student crushes, inciting and goading them on, and love affairs and complications would result. In all of this he was quite irrepressible. This obviously afforded him great satisfaction, and was not separate from his political methods but bound with them.

He played the role of a dissolute man, though he was not,

at least not to the degree he pretended. He was married to a pretty and good wife, and always complained that she never let him do anything. She would only smile at his promiscuous tales and his purposely unconcealed glances at every passing skirt. This philandering talk and pose were another inner need of this man. He could not live without a constant and lively *jeu d'esprit;* the deeds themselves were of less importance to him.

There was something quite suggestive about his staring blue eyes, set in a pale, sharp, and apparently tortured face, below a high wrinkled forehead and black, prematurely gray curly hair. He was the picture of a man rent asunder by an inner struggle, though he betrayed none of this by his actions. He was enmeshed in small-town affairs of every kind. Unlike others, however, he was never completely swallowed by them, but stood above it all, more lively than the rest, as though he held all the strings in his hands. In the monotonous life of the small town he did not know how to expend the energy of his ever taut and tireless spirit.

On the other hand, Zečević avoided quarrels. Once, he and another instructor, Miloje Dobrašinović, got into a political fight on the square. Peasants had to separate them. While they were arguing and shouting, the market place split into two factions, which surged like two waves of turgid and maddened water between which land had suddenly appeared. This was a shameful affair for the whole town, let alone the school.

Dobrašinović was an honorable man, a guileless character, but hasty and willful. He was an excellent teacher and a very considerate educator. Though his subject was mathematics, he taught it, too, with enthusiasm, and asked examination questions that helped the pupil along. He seemed more like a teacher of history or literature.

Certainly he was less to blame than Zečević for that clash; he was extremely sensitive about his pride, and Zečević could tease in an oblique but biting way.

Zečević was a strange combination of many talents, not one of which he particularly developed or exhibited. He made fun of romanticism of any kind, including even nationalism, and yet looked at reality and events himself with all the wishful thinking of the unrealistic romanticist. He could be extremely fair, yet he surrounded himself with cliques of students. He generously overlooked serious weaknesses, yet he invaded the most trifling thoughts and feelings of his pupils. He conformed to local traditions and ties, and at the same time he fought against primitiveness and encouraged more cultured behavior.

Zečević belonged to the Democratic party both before and after the dictatorship. Strangely enough, this inconsistent and self-contradictory man remained consistent not only under the dictatorship, but even after the war. He never changed his party or conviction. Devoid of any firm ideals or commitments, he nevertheless possessed a certain fixed fulcrum, invisible at first glance. Even he, who trifled with everyone and everything, was unaware of it. Time and certain conditions were necessary for this fixed fulcrum to become clearly visible to the outer world and to himself. He was not reticent about expressing his opinion of the postwar state of affairs, with a sharpness and clarity that astounded others. At the same time he sought a patronage which he himself recognized as being unlawful. For him there was no contradiction involved. He was simply a man with his own opinion of existing conditions, and who, in seeking patronage, was at the same time trying to survive under those conditions.

Extravagance and a debonair attitude and lack of consideration could once have been just a light *jeu d'esprit* in that time and place; consistency of views, however, meant a firmness of character which was exceptional in the postwar atmosphere, when fear and flattery increasingly became a way of life and human survival. All three—Mijović and Ivović and Zečević—were younger than Boričić and older than the greenhorns, the new graduates who filled the school as temporary special instructors. All three had finished high school before the war, had fought in the war, and finished the university afterward. The new instructors were postwar wonders who had finished two or three grades of high school in one year. The three instructors did not have any of the romantic ideals of Boričić's prewar generation. If they had ever had any, the war and life after the war snuffed them out. Yet these men had not sunk—at least not at that time and not completely—into the shallow humdrum of daily cares; they did not make themselves a career and a better life by elbowing, as the greenhorns did. Life had hurled them in various directions, according to their personal circumstances. All three were defined personalities, though a generation without firm common ideals.

The transfer of knowledge to others is also creative work. It saves those who do it from destruction; they endure and live in the realizations and intellectual ferment of those who are coming.

Both in Kolašin and in Berane, and probably elsewhere as well, the instructors who made the most painful impression were the Russian *émigrés*.

There were many of them, and they replaced one another frequently. Some were very expert and experienced; others were ignoramuses and derelicts. All were unhappy. They were at odds with everything and could not adapt at all. They were still too few to form their own community. They were usually torn into feuding groups among themselves, because they belonged to many disparate currents and ideologies, and because each was as exalted and spiritually exclusive toward the other as though the destiny of Russia, and even of mankind, depended on his personal fate and the acceptance of his own ideas. Their salaries were meager (few of them were professors by profession), and they were in constant dread of every superior, of the authorities, and even of the citizenry. Our food was strange to them—dry mutton, as hard as wood, salty cottage cheese, tangy cheeses, foul-smelling brandy, and mutton kebabs. Even our mountains were alien—not the gigantic and massive peaks of the Caucasus, but a sharp saw's edge that tore at the sky and obstructed the horizon. These Russians never climbed our mountains nor bathed in our cold rivers. Then there was our language—tongue-breaking. And our brusque people, who are not moved by tears. And our towns devoid of any intelligentsia, without any entertainment whatever—simple mountain villages with somber exercises on St. Sava's

Day and two religious thanksgiving days a year. This was not their homeland. And so they lived withdrawn in their little rooms, isolated and lost.

What they did and how they lived would be hard to say, but all of them, to a man, seemed like real eccentrics to the local folk.

Lieutenant-Colonel Kravchenko, in Kolašin, liked to beat pupils and pull their hair. Perhaps because he had to be given some sort of job, he taught sketching and gymnastics. Stout and muscle-bound gnome that he was, he did not seem to know what to do with his strength and military experience in this small mountain settlement in an alien land. He taught gymnastics with as much toughness and determination as if he were drilling soldiers on parade. He was heavy-handed and rough, yet appeared to be a good-natured and not very bright cavalry officer, which he probably was. He was too lazy for any other work, and had a sad yearning for his distant homeland. In the middle of the class hour he would begin to pace the floor—the planks swayed under his weight—and would sing drawn-out arias to himself. The pupils who learned these songs from him sang them without his sorrow and despair.

Alexander Malinovsky, thin and lanky, was so very shortsighted that it was incomprehensible how he ever told one pupil from another. The only explanation for his success was that he remembered their voices. He taught French, a language he had undoubtedly learned in childhood from his governess. Sloppy in appearance, he always chewed his nails, though his fingers were besmudged with ink. Since he was so nearsighted and careless, the pupils made him the target of various pranks. Meeting him on the street they would make a motion with their arms as though to take off their caps, and he would doff his hat. They would move the chair away from his desk, so that he learned to look carefully to make sure he would not be sitting on air. They even placed a row of bits of paper from the door to his desk, one for each step. He endured all of this without anger, even with a certain delight, as though he enjoyed the torture, being imprisoned and shriveled within himself behind those thick glasses.

Alexei Makhayev was also an instructor in Kolašin. Young, with ripe red lips, and more pretty than handsome, he walked with a brisk step, lost in thought and paying no attention to anyone. He was a good teacher of mathematics and quite an eccentric, like most of his compatriots. It was said that he slept with a cat, which kept his feet warm. He drank much goat's milk, because he was ill with tuberculosis. Unhappy and always on the verge of tears, he was also unhappily in love, so the story went, with Varvara Medenica. She was the one Russian woman there, the only lady whose beauty and manner were in the European style. Everyone in that backward mountain settlement who dreamed of something better or who remembered a more cultured existence was in love with her. Makhayev, however, was the unhappiest of them all. His love was not the real thing; it lacked that mad and desperate passion.

One Russian, Lebedev, did feel a real love, but for one of our women in Berane. Lebedev was still a young man, about thirty. He drank a good deal, like most of his countrymen, but rarely and rapidly, in moments of despair or resentment. He was a handsome man, all muscles, big blue eyes, and lithe though not tall. He liked to wear high boots, through which, as he walked, one could see the movement of his powerful leg muscles. Despite his beauty and strength, he would have remained unnoticed had he not been in love with Mrs. Popović, the most beautiful woman in town. And in many towns. A woman as beautiful as that never fades in the memory.

Her skin glistened with a mysterious luster from within. Her chestnut hair, which she wore long, rose in heavy and rich waves of old gold. As though aware of the radiance and power of her body, she liked to exhibit her bare shoulders and arms, which had dimples and curves that undulated one into another. She had a high waist, and was very trim in her high heels, with her small fine ankles. And her eyes were radiant, luminous, a golden green. The whole town shone with her beauty, and that beauty shone with her awareness of it. Her husband was hardly less beautiful than she. Their marriage was a happy one, though childless.

Did she encourage Lebedev and then retreat, or did he simply fall in love of his own accord? None knew; but the whole town was aware that his love was a desperate one. My friend Labud Labudović and I also knew. Labud, a cousin of Mrs. Popović, who lived with her, used to pilfer Lebedev's letters. It was all very strange, especially that Labud, the outstanding and best-behaved pupil in school, and also the shiest, should have joined in our deed. But we were in the pink of youth, read much, and everything new excited us terribly, though we felt shame for delving into other people's secrets. Letters as completely painful as Lebedev's could have been written only by a man who had lost his homeland and had never attained his love. Nothing in them was at all like what we had read in books, not even in those that spoke most openly of love. From these letters it was obvious that she had quite suddenly cut off their relationship, which had gone quite far, though apparently not to the end. Now he was begging her to see him once more, only once more, while her husband was away on a business trip. We did not get to read her letters, but it was evident from his that she was shrewdly and stubbornly resisting him.

Lebedev wrote quite openly about everything. One night, quite desperate, he rushed to a tavern, got drunk, and then—she knew where—to a certain singer. Another time he stood outside her house till late at night to see her shadow on the window shade as she was undressing. After that letter she never forgot to pull down the thick shade as well. She became more and more distant. But the cruelest of all was the letter in which he told how he went one autumn night into the grassy meadows to kill himself. There were many stars in the sky and everything seemed to him wretched and mean, he and that love of his; even she, the beloved one, appeared petty and selfish. It was not hope that she would be his that kept him from suicide. No, on that night even that seemed trifling and unimportant, beneath alien peaks and alien stars, on the damp grass. He spared his own life to spite himself, to torture himself. Our men do not love like that, nor can they.

His letters were short, full of unfinished sentences and disjointed words. One could hardly gather their meaning. The lady found them to be clear, but all in vain for Lebedev. He left soon after, fled the city, perhaps to find a new unhappy love on another foreign strand.

It would be hard to find a more good-natured man than Professor Shcherba. He never cared for teaching, nor for anything else except liquor—what kind did not matter, as long as it made him drunk. He would usually get drunk in town at dusk, then weave his way home and sleep it off. He was sunk in liquor and in a dark despair, utterly. With a sparse beard and a wrinkled brow, he was short and sluggish. He peered through his glasses with the worried eyes of a father of a large family. He loved everything, both people and things, with an obtuse incurable love. He never gave poor grades, except perhaps as a warning that one had to learn something, after all.

Another Russian we called Gallop because of his swift, almost running walk; he could never have been a teacher anywhere else but here, because of our dire scarcity of teachers. Small, swarthy, thin, he was seriously ill with weak nerves. He would be seized with such fits of weeping and rage that he would beat his pupils and then kiss and slobber over them.

Our sixth year was distinguished for its *esprit de corps*, established even in the fifth year in the struggle against the

teachers. Several of us led the whole class in a strike against
Gallop and his antics. A newcomer, named Branko Zogović,
refused to join us, perhaps because Gallop loved him madly,
perhaps because he was inclined to get out of risky under-
takings even when he approved of them. The strike succeeded anyway—and Gallop was withdrawn. Gallop was a
completely lost man, so utterly befuddled and wrapped up
in his own world that he saw nothing but his own unhappiness.

The Russian *émigrés* did not stay long. The community
did not tolerate them, nor they the community. Restlessness drove them to go their way.

One of them, however, remained from 1921 or 1922
until his death in the 1930's. He was named Krestelevsky,
or something like that. But the town and the school called
him Baldy, for he had not a single hair on his head. Most of
the Russian *émigrés* looked like eccentrics because they
were in an alien community and crushed by longing for
their homeland. Krestelevsky, however, was an eccentric
in his own right. He did not mix even with his compatriots.

He was already over seventy. One could tell by his bearing that he had been an officer all his life; he was erect, stiff,
with a heavy tread, always dressed in a gray army great-
coat buttoned to his Adam's apple. He had come in that
coat, and was buried in it. He never joked, never smiled.
He was the only Russian not subject to those sudden
seizures of tenderness and wrath which are so frequent
among them, especially when they are far away from the
homeland they love with an unabated and intransigent sadness. He always had an icy expression on his face, as though
he were without heart or soul. He acted the same way
toward the pupils. We felt that, in his eyes, we were im-
personal, arranged in rows, not even like soldiers in the
army ranks in his boundless country, but more like bullets
in the barrel of a revolver. His outbursts of anger were like
that, too, military. He would suddenly bawl out in wrath

and strike with his boot, and then just as suddenly calm down as though nothing had happened. His most frequent curse was the Russian *sukin sin,* son of a bitch, but only when he flared up. The cursing and the bawling and all the rest were for him an inevitable procedure, necessary wherever there were human beings, who were born to be drilled.

He taught chemistry, and other subjects on the side. It was evident that his field was something quite different, for he had learned by heart the entire chemistry textbook, which was written, of course, in Serbian. He would recite whole passages from it exactly, without omitting a single word; and when he got angry, he would even cite the page where a certain section could be found, and the punctuation as well. How could a man of his years, unacquainted with the language (for he knew Serbian very badly, and drilled and spoke more in Russian) learn a subject that was not in his field and in a foreign tongue? Was he afraid of losing his job, and therefore wished to master the subject? Or was this conscientiousness? Or did he use this means to kill time, this lonely man in a strange land?

He lived alone up above the town in a little room. There he cooked and washed by himself, never going anywhere for years except to school or to shop. He was extraordinarily clean and neat. He never drank, and lived on tea and toast. From his pitifully meager salary he sent help regularly to a kinswoman. The lamp in his room would shine late into the night. What was he doing, this lonely old man who was thrifty with everything apparently except kerosene? Could he be going over his chemistry book, so as not to forget in his old age? Or was he reading the Bible? Nobody knew.

All of us pupils believed that he was a cruel and heartless old man who hated us, the town, and himself, and that if he could, he would crush the whole world under his boot just to enforce quiet and order. But it was not so. He, too, had a soul.

In the fourth year of high school we still did not know much about politics, and were even less able to delve into the intimate aspects of human life. In the sixth year, however, things were different. Once, when Krestelevsky substituted for another instructor, we began to ask him questions. Did he have a family? Why was he alone? Then the iron old man began to weep softly. He was ashamed of his pain, so he turned to the wall. But we could see the tears sliding down his whiskers and dripping on the dusty floor. We were all shaken, and the girls began to wail. His family had been killed in the revolution.

In the beginning, Krestelevsky was ridiculed, then hated, and finally respected, because of his ascetic life, his conscientiousness and strictness, his fairness in everything, and his unassuming nature, and because of the help he sent, at the price of going hungry, to someone far away in the world. He was like a living saint. When he died the whole town mourned him, though admittedly with a brief and hardly noticeable mourning, as though some rare plant had withered in one of the town gardens on the outskirts.

This sorrow and inability to adapt oneself after being uprooted exists not only among humans—as was so evident among these *émigrés*—but also among beasts. It is strange to tell, but the Austrian cows that my father obtained from the state as war reparations reminded one of the Russian *émigrés*. They were sold rather cheaply, to improve the breed of our own puny beef cattle, and so Father bought two. He thought he had done enough in buying the purebred cattle, that they did not require any special attention—food and care—other than what the Montenegrin scrub cattle were getting. It was soon evident, however, that they needed a different kind of care. But since such care did not, and could not, be, both cows, Jelulja and Bjelulja, were obviously unhappy in this land of rocky crevices and hunger in which everybody hated them: the other cattle hated them because they were so awfully big and slow, and the people because they did not breed easily, gave too little milk, ate much hay, and were never full. Big, awkward, and always hungry, they would get into crevices where even our cattle did not dare go, and then they could not escape. They would wander off from the rest of the herd and get lost, and we would have to look for them at night. They did not even know enough to be afraid of the wild animals.

Jelulja was calm and tame, Bjelulja was wicked and spiteful, very thin, with a hard and diseased udder, always ready to take a swipe with her horns. Jelulja was such a good-natured beast that my younger brother, Milivoje, and I rode her across the Tara to avoid having to take off our shoes. She soon got accustomed to this and would wait for us at the bank herself.

Our bulls were too small for these cows, and their progeny was poor and never lasted. No good ever came of them, just sorrow and trouble. But animals try to run away from sorrow. Men wallow in it, not seeking balm for their ills.

It is not true that one's homeland is wherever it is good. Man is born only once and in one place. There is only one homeland.

Besides the serious instructors, whether strict or gentle, there was another kind in Berane, as everywhere else, and perhaps the majority—those who were not at all suited to their calling. The school could not do without them; someone had to teach. Nor could they do without the school; somehow they had to make a living. Most of them were university students who lacked their degree.

George was one of these instructors. He taught history. He knew all the dates, names, and places, but one could never learn from him the true substance of the historical process. He was convinced that he was an excellent stylist and speaker, though his lectures were disorganized and bombastic. We students, in the seventh and eighth grades, sensed his weak point and, to make fun of him, would come up to him with the flattering compliment that he spoke more picturesquely than Carlyle. Only two or three of us had ever read that author, but, through George, we all knew that Carlyle was his ideal.

George lectured something like this: "At that moment, with the rage of a wildcat, Napoleon leaped on his horse, drew his sword, and shouted, 'Attack, follow me!' And the obedient legions followed. . . . Robespierre leaped to the rostrum as though scalded. . . . Cicero spoke out his famous words as loudly as he could, so that the Senate was left thunderstruck. . . . All around there reigned a foreboding silence while the Persian army entered the gap, not suspecting that black Death lay in wait for them.

. . . Mortally wounded, Caesar declared in a sobbing
voice . . ."

This style of lecturing, accompanied by the waving of
arms and spraying of spittle, was extremely funny. We
took to imitating him, even in front of him. He would not
always lecture like that, but we would have to flatter him
by telling him how much his Carlylean style inspired us
and carried us away, and how his lectures, with all respect
to the other instructors, were the most interesting. All of
us knew, to a man, that this was all in fun. Yet we were
united in a plot against the instructors, particularly those
like George. Despite these lectures of his, and the fact that
he did not yet have his degree, one cannot say that he did
not know his subject, though he knew everything by rote.

But George had even greater faults. He was convinced
that he was handsome and witty, though he was neither. On
the contrary, he was as ugly as sin, gawky, his face eroded
with pockmarks, a nose like a sickle, and bulging lifeless
eyes; he was quite dull-witted, though not slow.

He dressed in tasteless and loud clothes, wearing yellow
pointed shoes and gawdy ties on every occasion. He would
ride into town from a nearby village where he lived with
his family, not because of the distance, but just to show off.
Conspicuously and awkwardly, he would set his horse to
prancing as he passed the windows of the better-known
girls. In vain he made eyes at the girls in town, and at the
prettiest ones. The pupils knew this, and those in the upper
grades took advantage of him. Most active in this game was
Vule, a scrawny peasant kid with a stupid look. He would
approach George with especial *savoir-faire*, playing the
part of a peasant simpleton. Once, Vule informed him the
whole town was saying that he had fifty suits and that he
had paid a thousand dinars to have each one made. George
was obviously flattered and, of course, did not deny anything, but merely observed modestly that this was somewhat exaggerated. Not long after, Vule again descended.

"One hears, sir—but please do not be offended at my mentioning it—that you had one hundred and fifty love affairs in Belgrade." All aflush from the unexpected pleasure, George again made a modest denial. "No, no, what nonsense. All of this is exaggerated, quite exaggerated." But he pronounced this denial in such a tone that everyone could have concluded that Vule was not only not exaggerating, but that he had rather shot under the mark. And so it was with everything about George—how he was the best student and knew more than the professors, and how they envied him. And Vule also brought him alleged greetings and messages from the town beauties.

George loved our class, and used to say that whenever he came to us he felt it was a real spiritual treat. Maybe there was some truth in this. But we, too, had a treat with him. The relentless town made fun of his awkwardness and his airs, behind which, they knew, hid an unpolished peasant. They never forgot how he had been shoed. Before going to the university after finishing high school, George went to buy a pair of shoes, and the shoemakers, while taking his measurements, ordered him to lie down, lifted his feet, and pounded his soles with their hammers. He was not such a bad match for marriage, yet this story alone scared off the better girls. Alone in his make-believe world, aware that he was being ridiculed, failing, despite his stubborn efforts, the examinations in Belgrade, it was natural that he should have felt misunderstood, and so he came to believe that only in us could he find some warmth and the kind of noble understanding that only unspoiled young hearts could offer. But even this was make-believe with him. Our warmth was staged. We made fun of him by flattering him, and he believed that we liked him.

If George was a tragicomic figure, Luka was a comical one. Squat and fat, he neither knew anything nor studied anything. His knowledge of his own subject was less than that of any one of his pupils. He never lectured, but would only tell us to read from page so-and-so to page such-and-such. In the upper grades he hardly ever drilled the students, or only very superficially, just enough so that no one could say he did not. In giving grades, he looked at the student as a whole, and acted accordingly, seeing to it that nobody received a failing grade. He could judge by his own case how much happier human beings would be if there were no failing grades. He regarded it as very important that every student should be a good comrade, an all-around man, and the like, that he should like a good time and not be a bookworm. Nothing else mattered. Our class hours with him were unbearable, and for him killing. Frequently, he would take us outside, into nature, supposedly to explain something to us on the spot, but he never made the connection. His sessions in the classroom were spent in disjointed discussions about everything under the sun, hardly ever about his subject, or only incidentally so.

On one occasion a certain student, another big prankster, brought a bomb to class. Several of us got together and told Luka about it. He immediately agreed to our proposal to spend his hour by going to the Lim River to kill fish with the bomb. No sooner said than done. The girls had to remain a distance away from the pool. It would have been embarrassing for them, because we had to strip naked to bring in the dead fish.

He joked with the girls, and in an improper and unseemly manner. Dunja Vlahović was already a mature and quite pretty girl. He caught sight of an apple on her desk and tried to take it away from her. She felt she was within her rights not to let him have it. Only when she realized that he was not interested in the apple but in grappling with her did she retreat in confusion.

On market day he would get drunk, but not so much that, after heated discussions with the peasants, he could not walk to his village under his own steam. The town urchins would shout after him, and he would shake his stick and curse their gypsy tripe-eating mothers. He, like the peasants, considered the townsfolk like gypsies and believed that they ate only intestines, tripe, hence his choice of invective.

He was a great one for hopping on band wagons. First he was for the ruling party, the Radicals, and later for the dictatorship. It was not very hard for him to get into a quarrel, and even a fight, over the elections and his party, though he was not at all pugnacious but, like every peasant, stubborn and unreasonable when it concerned something close to him.

I was perhaps the only student who had ever had any unpleasantness with him, and in my last year at that. During his class hour I buried my head in my hands, probably because of a headache, and lay down on the bench. He demanded, "Why have you buried your head like a hog in the slop pail?"

To this I answered, "You ought to express yourself in a manner more befitting a teacher."

He reddened completely but said nothing. The next time he called on me to recite he gave me a failing grade. True, I did not know the subject. Neither did the others, except for the few who learned everything by rote. There was an unwritten rule in the school that, except in special cases, failing grades were never given in subjects that were not included in the final examinations. Apparently I was a special case. I knew that it was useless to study the subject now, for nothing depended on my knowledge any more. But I had some good friends in another section of the class. They got Luka to reveal to them the grades at the end of the year. He told them that I alone had a failing grade. Feigning astonishment, the whole section fell into an uproar and tried to prove to him how senseless this was, what a good fellow I was, a bit touchy, perhaps, and impetuous, but . . . and so on. He relented and gave me a passing grade on the spot.

The students did not make fun of Luka or play tricks on him. There was no need or occasion for this. Not that he would have caught on, but he was such an open and above-board character that he had no pretensions whatever, and there was no cause for pranks. Our relations with him were simple, as he himself was. They were the relations between peasant neighbors, except that he was a bit older than we and had authority over us.

His own brother was our classmate and his direct opposite—like those two kids of one goat; according to the folk tale, from the skin of one of them they made a war drum, and from the skin of the other a binding for a Bible. He was a good student, with a special gift for history, attentive and conscientious in everything. He, too, noticed the strange pedagogical qualities of his brother and, being a bit ashamed about this, tried to influence him, but he was also weak in this regard, and a brother is a brother.

Nor did we make fun of another instructor, Milutin, for entirely different reasons, even though he was a bit queer. He never allowed the slightest intimacy between himself and the pupils.

Milutin undoubtedly had a vast knowledge of his field, and it reeked of mothballs and smacked of hairsplitting. He had studied in Germany and relentlessly read thick German tomes. But he read these books in a conspicuous manner—on the window sill, even in the street. He had a stern look, wore pince-nez, and was all in gray. He considered himself an unrecognized genius with a rare profound intellect. He had some reason for this opinion—his study abroad, and the fact that he was from a renowned chieftain's family.

Apparently the only approach to use in tricking him was a profound subject, naturally a German one, for nothing else would do. We had to be oblique. For example, we had only to declare that the greatest philosopher, poet, or the like, was a Frenchman or an Englishman. Milutin would immediately undertake to demonstrate that this was not so,

but a German, German. He was an incurable Germano-
phile—except when Serbs were involved. The Serbs were
even better than the Germans. He himself was a case in
point. While he was a student in Germany, some German
students, *Burschen*—and there were ten of them!—came up
to him in a tavern and called him a *Serbisches Schwein*, a
Serbian swine, for which he beat them all to a pulp. The
newspapers, German newspapers at that, were full of news
the next day about the brave Serb who . . . He told the
story very animatedly, showing how he punched one,
flipped another one over, and so on, all the while straining
his weak muscles and caved-in chest and coughing and
spitting in his handkerchief. He had had some ribs removed
because of tuberculosis.

Nothing pleased him as much as to tell about Professor
Haeckel, the well-known German naturalist, whose student
he was, but also, as he liked to boast, whose friend he was.
Milutin described that friendship as being one between
equals; indeed, he was doing Haeckel a favor. He happened
to be reading the man's work one day in the park. A
dignified old man came by and asked him what he was reading. "Haeckel," answered Milutin.

"I am Haeckel," the old man introduced himself.

"I am Milutin," our sage replied.

In the telling, Milutin pronounced his own name with
such dignity and self-esteem that one could hardly imagine
but that Haeckel already knew of him. A conversation with
Haeckel ensued. Milutin even made certain criticisms—yes,
yes, criticisms—some of which the old savant had to accept.

According to him, the two had corresponded until recently. But not regularly. "My fault," Milutin would say.
"I hate to write letters."

This relationship with Haeckel had its sentimental side as
well, one which we could guess—but only guess. Milutin
told us how frequently he used to visit Haeckel. We would
send out a feeler: "Did Haeckel have a family?"

"Only a daughter" was the answer.

We dared to ask, "Was she pretty?"

"No, she was not pretty" was the answer.

We dared not inquire further, for Milutin did not tolerate any intimacies and familiarity on the part of the pupils. Besides, he had already said all that was needed for our imagination.

We did not know what Haeckel's daughter looked like, but Milutin's wife was truly not pretty. She was a stout and fat lady who seemed to have siphoned off all of the imaginary strength of her husband. Yet in every other way she acted like her husband and imitated him in everything. She, too, believed he was a genius, and maintained an extremely stately bearing, moving about as slowly as a ship and sitting motionless like a statue.

Once, as though by chance, I brought a colored picture of Schubert to class. He noticed it and asked me what I had there.

"Wagner," I replied.

"Not Wagner, but Schubert," he observed. "And do you know who he is?"

I pretended that I did not exactly know. "A player, or something like that."

He began to talk to us, though not much of Schubert, about whom he obviously did not know very much, but about Wagner's operas, about their contents and brilliant scenery, the choruses, duels, castles, knights, and fairies, and how skillfully and picturesquely the Germans did this—as though these things were the essence of Wagner and his operas.

We could similarly use one of his periods by getting him wound up on the subject of German literature, and thus once more escape being examined or having to listen to a lecture. For example, someone would raise the question of which was the greater, Goethe or Njegoš? Something of the sort could never pass without a prolonged discussion on

his part, with the stern observation that the question itself
showed how poorly, how very poorly and superficially our
instructors in literature taught their subject, and he regretted that he had to carry out the tasks of other people.
But the greatest and most successful topic of all was German philosophy. Hegel, Kant, Fichte, and Leibniz—he
talked of them as he did of Haeckel, as good and close
friends.

Milutin was not strict, but one had to know his subject,
at least to some extent. Good grades were not easy to get
from him. Indeed, something of the German spirit did stick
with him—conscientiousness and exactness, of which, by
the way, he was inordinately proud. This man did not put
on airs simply to dazzle us and the community. He deeply
believed that he was a genius on the order of Haeckel or
Kant, and did not care much that he was not understood.
This was evident in his every gesture and whole bearing.
He never took a walk or sat in a café without appearing
deeply engrossed, with a wrinkled brow, and even a lost
look. He looked like this on purpose. This did not mean he
was only pretending to think. He simply walked slowly so
he could think better, and he thought as though with the
knowledge that he was doing some very important work.

I have never seen a man who regarded his own dignity
with such self-confidence. This man could never be unhappy. By dint of sheer will—and he believed, anyway, that
the will was the most important thing in the universe—he
constructed a world about his own greatness, and to the
degree that others did not share his view, he believed in it
all the more firmly. There was much in him of the local
nobleman who thinks he is a philosopher, while his noble
lineage and conviction enable him to play his imaginary
role and to rejoice that he can be above the herd and alone,
in a world that exists only in his own head.

Tragedy was not absent from the fate of our local poet,
sculptor, and painter, Limski. He was a man of little

schooling, almost self-taught. He was convinced that he was a sculptor, had some reputation as a poet, and lived for money. He came from dire poverty and went out into the world to fill his belly. He had learned sculpturing somewhere or other, and the war found him with the Serbian army on the Salonica front. It was there that he began to write poetry. Though unschooled and without any gift to write significant poetry, he nevertheless felt, better than many such poets, the sad yearning of the simple and unimportant soldier in a foreign land for his homeland, his home and wife, his cattle, stream, fields, and orchards. This was not really poetry but the simple expression in rhyme of the sorrows of ordinary soldiers. His little poems were eagerly read and easy to remember, and just as easy to forget.

Apparently, to be a poet in a small town was to be ridiculed. Even without this, Limski gave cause for ridicule. His very appearance was unpoetic—bony, red faced, with trimmed gray mustaches. He looked more like a petty clerk than the kind of pale and sad-faced small-town stereotype of the poet. True, he did wear his hair long. But it hardly looked poetic—more like plaited whips. He was a great miser, the sort that orders a glass of water and a toothpick in the coffeehouse. He lent money at usurious rates, mostly to his own fellow villagers and always through a third party. Though he tried to look dapper and original, everything he wore was cheap and loud. His miserliness, masked by extravagance, was evident also in the way his house was built. It was narrow, but it had everything the more sumptuous ones had—a balcony and a decorative façade, and on the balcony were three of his sculptures: of Vuk Karadžić, Dositej Obradović, and Bishop Strossmayer. He would not have Njegoš. He was jealous of him as a poet. As he made more money, he added to the house, first one wing, then another. Then the balcony was enlarged. But he never added any statues. The enlargement of the house lasted for years, and Limski changed with it. He gradually got fat, added a bay window, and his bony face became even redder.

He also brought a wife into his house, which was by then a villa with the patriotic name of Corfù, the island to which the Serbian army retreated on leaving the homeland in 1915. His wife was pale, almost dead white, with heavy black hair and dark circles around her eyes. She was the one who really looked poetic, though she was merely a sickly woman who was bored by everything in that forsaken town. Limski treated his wife with attentiveness and respect in public. But the neighbors could hear at night how he shouted at her crudely, and maybe even beat her. The villa resounded with unpoetic phrases and the crashing of furniture, with cries, curses, and moans.

Limski had a talent for cashing in on everything. He looked upon his patriotism and his poetry as a way of obtaining various sinecures and grants, as a man who has deserved well of his country. He used to send his books and poems to distinguished men, to whom he would dedicate his works. He succeeded in having his inscription placed on the triumphal arch in Berane in September 1925, on the occasion of King Alexander's visit. The verses alluded to the fact that the King's forebears were from that region. They read:

> *The native soil of a glorious vine,*
> *The cradle of thy Grandsire, thine,*
> *For ages has awaited thee,*
> *King of glory, victory!*

People stood that day in two lines along the main street. We schoolchildren were closest to the ropes that were stretched out along the whole length of the street. Flowers were put into our hands and we were told to throw them on the street before the royal pair while crying: Long live the King, long live the Queen! We all knew that the King was of small stature, but he was so small and thin that we were completely disillusioned. He walked, beside the already corpulent Queen, with a step that was too long and strenuous for his thin and frail body. There was a good deal of sincere enthusiasm, especially among the old soldiers and the youth. The old soldiers saw passing before them their dream come to life, albeit a scrawny and unprepossessing one. The old men recalled their battles and wounds and fell into a childish emotion, while the young people shouted for something they believed would yet come. Even though, as usually happens, the moment of meeting with the ruler, so long awaited, turned out to be rather commoner and more staged than was suspected, still everyone was full of drunken enthusiasm and abandon.

While waiting for the King, Limski nervously paced up and down, frequently passing his bony fingers through the whiplashes of his graying hair. From time to time he would smile knowingly. He was preparing to say, to recite, something great. When the King appeared, shouting was heard at the town gate. Limski stepped aggressively to the triumphal arch, under his verses. But somebody from the welcoming committee shooed him away. That was the first blow. The second was struck by the King himself. Quick and impatient as he was, he did not even read the all too long inscription on the arch, much less ask who the author was.

Though he himself gave rise to ridicule and abasement, Limski was really the victim of small-town boredom. They made fun of him and invented spiteful pranks which had no connection with his failings. They ridiculed him for being a poet and an artist, and not for the things he created. They mocked him for the sake of mockery, teased him and played tricks on him.

Despite all this, he had something inside himself, something basic, which no amount of ridicule, or even his own weaknesses, was able to dislodge: he knew how to en-

courage the poor pupil and to keep him from becoming dis-
couraged. The subject he taught in the school—art—was not
important, nor was his position among the other instructors.
But among the pupils he meant much more. He possessed
warmth and the skill to attract and to give courage to dis-
couraged students. One could discern in him the poor boy
who broke out of a backward village into the big wide
world, with a determination and fire which became a part
of him and which no failure could extinguish. This was the
very trait his mockers refused to recognize. In their mockery and invention of his characteristics, they lost sight of
the man in his entirety, and even of the man they failed to
imagine.

For every man, and for every group as well, the most
real thing is the very thing that others most frequently attack. Only the attack is real.

The high school in Kolašin was not as crowded with so many students as the one in Berane. Lack of space, an insufficient number of good instructors, favoritism of every kind would have made teaching difficult even without the lack of discipline and the wildness of the students.

The students had reason to complain about many of their instructors. The instructors had even more reason to complain about their students. It was really a perfect match—generally incompetent instructors and headstrong students. The fifth grade, in which I was enrolled in the academic year 1924–1925, was certainly the outstanding class for lack of discipline and most responsible for the chaos that reigned in the school. With what trouble even a little bit of knowledge was gained! For lack of classrooms and teachers, my fifth grade, in which there were sixty-four pupils, was not divided, as was usual, into two sections. Had that mob been composed of the best-behaved students, and if it had had enough excellent teachers, it still would have given vent to its savage and irrepressible impulses under those conditions and become a source of disorder and chicanery of every variety.

The class immediately sensed its power and became a solid group united in a common cause against the instructors as well as by the pranks into which the majority were gradually drawn. There are always pupils in every class who, when something happens, are not able to withstand the pressure of the instructor but betray the culprits.

There were such weaklings in our class as well. But they lived under such pressure, in such terror of the majority, that even they had to keep still when the going got rough. The great differences in sizes and ages—for among us there were already young men—made it all the harder for the school authorities, but this also served to instill terror in the weaklings and traitors, who were in danger of being not only ostracized but beaten up. Good excuses for this procedure were never hard to find.

It was easy to create disorder. All sixty students had to do was to begin, by spontaneous agreement, to shuffle their feet under the desks, or to cough and sneeze as if suffering from colds. There were other ways, too. Everyone would begin to mumble, as though going over lessons to themselves, or all would whisper to the pupil who had been called upon to recite. All those mouths whispering created an uproar.

To be sure, we picked our instructors and the methods we would employ against each. There were also those in whose periods we were quiet and orderly. But even that was put to good use, as proof for our home-room teacher or principal of how unfounded or exaggerated were the charges of those instructors against whom we had sinned. We declared our innocence unanimously and so convincingly, with the strength that only a mob can have, that the principal and home-room teachers frequently wavered in their own convictions or retreated, unable to discover the ringleaders of the coup. There were not many of these ringleaders, about fifteen students in all. I was among them, distinguished more for my ability to dream up new pranks than for my participation in their execution.

We found it easy to bedevil the non-Montenegrin instructors with disturbances and pranks. In this our perseverance and ingenuity were inexhaustible. The women had especial trouble with us, particularly since all of them were foreigners. But the Montenegrins, with their dread severity, were not to be trifled with. They found it easy to keep track of our habits and skills. We were just the pupils for them. And they were the teachers for us.

From year to year we never had regular instruction in any single foreign language, though we studied three. The instructors in these subjects replaced one another all the time. Each instructor introduced his own method and generally began from the beginning. His successor never got much beyond that beginning. French we had from the second year. Instruction in the German language, which had just been started, might have developed successfully with a new instructor had the class not been the kind it was, but we made this utterly impossible. The new instructor was Mrs. Lazović, a native of one of the Baltic countries. Everybody thought, though, that she was a Russian, because her homeland had been a part of the Russian Empire until the war. She was extremely educated and had such a command of the Serbian language that one could barely tell that she was a foreigner. Frail, slender, with blue eyes, she seemed to be a miniature in pastel. She held her pen very tightly, so one thought her spindly fingers were going to break; and her heels were very high, so that one feared she might fall and break into pieces. Though she was nervous, like most frail people, she was determined, not so much to maintain discipline as to get us to like to study German. However, the more zealous she was in her effort, the more stubborn we became. She began to teach us songs in chorus, which we turned into a howling bedlam. We also made impossible the conversation method. The written assignments we simply copied from one another. We organized a general stomping of feet. The principal and home-room teacher burst in several times, conducted investigations, punished anyone that even looked guilty, expelled two pupils, and then had to reinstate them upon the intervention of some citizens and instructors. As a matter of fact, that pair was no more guilty than a goodly

half of us. The determined and nervous Mrs. Lazović per-
sisted, however, until her judgment day came.

   During a recess we noticed a stray dog wandering around
the school. The next hour was to be with her, so we trapped
the dog under the rostrum on which stood the desk. Mrs.
Lazović mounted the rostrum and made some entries in
the class record. In the room there reigned a silence that
amazed her. Perhaps she might have thought that the class
had begun to mend its ways. We were taut with expecta-
tion. The minutes went by, but that cur kept still, as if
determined to betray our hopes. We began to lose hope
and to forget our prisoner. But then somebody began to
kick up a fuss, and Mrs. Lazović, as usual, began to pound
sharply with her heels. The dog finally began to bark.
Pandemonium broke out in the class. Mrs. Lazović grabbed
the grade book and rushed out.

   We knew that Principal Boričić would come flying in
immediately. We quickly lifted the desk and the rostrum,
dragged out the dog, and lowered him out the window.
The principal rushed in, this time genuinely enraged. "Ha!
Now you're in for it! At last we shall have a reckoning!"
He announced right off that he would expel every tenth
pupil if we did not reveal the ringleaders, and he began to
count off. The entire class began to assure him that Mrs.
Lazović had been mistaken, that there was no dog under
the desk but under the window. The principal cast a
glance outside—a good sign that he was in doubt and ready
to weaken—and he noticed the dog in the schoolyard. The
entire class sensed that the principal was wavering and
began to shout assurances. He did not expel anyone, but,
being unconvinced, he did not wish to let us go unpunished.
He began to slap every tenth pupil, just like executions in
time of war. But he had slapped only three when, on our
renewed entreaties, he called off the punishment and announced that from now on he would instruct us in Ger-

man. He taught us only a short while, though, for he was occupied with other tasks.

The case of Mrs. Lazović, who then stopped teaching our class, and later left the school altogether, was by no means an isolated one. A certain Serbian and his wife remained a month and simply fled. We could trifle with Mrs. Lazović, but it was impossible to do so with Mrs. Ugrić, a native German, for the simple reason that she was able to stand anything. She came to teach German later, after the fifth grade had been divided into two sections; it was extremely difficult to keep under control a mob that had already become accustomed to having its own way and felt protected by all kinds of external pressures—clan, family, and even party ties.

Mrs. Ugrić was a rather stout woman, with a mottled skin, placid yellow eyes, and extraordinarily distended nostrils, whose outer rims were turned up so that the dividing section underneath was quite visible and exposed from all sides. This made the expression on her face similar to those good-natured, easily bewildered and startled animals like the deer and the cow.

The rostrum in our section—we were then in the seventh grade—had become rotten. In one spot a board had weakened just enough so that the pressure from the leg of the chair would send it crashing. No amount of adjusting the chair on our part could ever fool Mrs. Ugrić. She would always carefully look before and around her, as though feeling with her eyes, and only after moving the chair would she sit down securely behind the desk. Nor did any of our other tricks work. Once we even cut open the sponge, poured ink into the center, and sewed it together again. But she did not soil her fingers. As though sniffing out the situation, she carefully prodded the sponge and, without getting upset in the least, sent the monitor of the day to wash it out.

Mrs. Ugrić paid no attention at all to noise, whispering, and the like. It almost seemed as if this pleased her. On the other hand, our behavior was such that one could hardly detect that any break had intervened between recess and the beginning of the class hour, except perhaps that we remained at our desks and that individual shouts were now replaced by a buzzing monotone. That buzzing was seemingly the result of an agreement between us and her, and even between her and the whole school. Amid that monotonous hum she would call on a pupil to recite, ask him to come closer, and then would put questions to him. It was very hard to drill under the circumstances, but she did not get upset, apparently reconciled to the idea that in this land and in such a school no one could teach these pupils German, or anything else, for that matter. All education here was simply some sort of solemn observance, like the ceremonial in church, which is observed even if the faithful, and perhaps the priests as well, do not themselves understand the liturgy.

It appeared that she was so calm and indifferent to everything that nothing could stir her passive heart and disturb the established rounds of a life accustomed to strict order and cleanliness. Insofar as anyone could see, she was also like that in her family. The whole town used to say of her, "A real German." They meant to say by this that her nation was known for its order and precision, but also by its unfeeling patience. Actually, she was not like that at all. Beneath that calm there was hidden an unusually sensitive nature. Perhaps one might offer as proof the fact that she later suffered a nervous breakdown as a result of her almost sick tenderness toward her own children whenever they got into any difficulty. Her sensitivity and tenderness were immediately discernible in her special subtle understanding for the problems of pupils in this wild and alien community, something of which not even our own people were capable, at least not in that measure. As though with the aid of some additional sense, she would ferret out the fact that a certain pupil could not study because he lived so far away, that another had no suitable living quarters, that still another had too many chores to do at home. This comprehension of hers was all the more wonderful and inexplicable because pupils in her native land certainly never had such difficulties. She never pried into the pupils' troubles. It was enough for her to hear two or three words to grasp the situation. She would look at the pupil with her always calm and wondering eyes and put off the questioning to another time when the pupil was better prepared. She almost never gave failing grades, realizing probably that under those conditions this would not help. She was probably adapting herself to the general philosophy that it was best for as many as possible of these dirty and savage little peasant children to finish school and find their own way in life, and if any of them should ever need a foreign tongue, they would take care to learn it soon enough. When they left the school, these boys, such as they were, would, nevertheless, be more cultured than their fathers. Civilization here must begin with them, ready or not.

And indeed, not one of us left the school with a knowledge of any foreign language sufficient for any use, even for our further studies. Not even the best, who devotedly studied a language, went much beyond a very stilted and dead book learning, of not much use, in the final analysis, except to ensure good grades. Actually, we did not have a single instructor for any longer period of time whose field was the language he or she was teaching. The same frequently applied to other subjects as well. The only exceptions were mathematics, physics, and the national language, where the knowledge was on a somewhat higher plane. The instructors, most of whom were without their degrees, were themselves dependent on crass considerations, and so they, in turn, promoted crass considerations.

The school was one tight and snarled knot of interests and influences of every kind save academic and pedagogical. It was more of a struggle than an education. A struggle against everyone and everything, least of all to gain learning.

It is strange, but with the advent of the dictatorship of January 6, 1929, conditions in the school took a turn for the better. If nothing else, all kinds of conflicting interests abated, and administrative order and discipline were strengthened. At least now all the instructors were dependent on one center of power and not on many, as before.

Somewhere about that time a new principal was appointed, Dr. Ante Mišura, a Dalmatian, a very serious man, not subject to various influences. He was one of those inconspicuous but diligent workers whose unseen hand is soon felt. He was independent of both local and partisan influences, and because his knowledge, especially of Latin, was considerable and solid, he did not have to take a back seat in this respect to anyone. He was obviously a man accustomed to order, and so both instructors and students took an immediate dislike to him. But nobody dared rise up against him.

This is not to say that the dictatorship sent Mišura, nor that he found in it his opportunity. Mišura would have established order in the school anyway, regardless of the political tendencies of the dictatorship. These tendencies began to be felt from the first days of the dictatorship, through an emphasis on a nationalistic and monarchistic spirit in the instruction and in the elimination of instructors who appeared to be possibly unreliable. In any event, the time was ripe for an improvement in the school, if for no other reason than the fact that the prevailing conditions had become unbearable both for the pupils and the teachers themselves.

Later conflicts in the school were of a different kind—a

political tussle between the regime and its supporters in the school and the leftist-oriented students. But that period was hardly noticeable while I was a pupil. The clash in my time was different; it was between savagery and a forcible order. It was a prelude to the later struggle. One led into the other.

As is generally the case, one form develops from another form, so that it seems as if the later one completely pushes back its predecessor.

In this land one believed more in fairies, witches, and vampires then in any idealized and inscrutable Christian or any
other god. God was only a phantom who was good to the
good and bad to the bad. Christ and Mary were not much
more real here than were good spirits. The cross was a
good omen for driving away evil spirits and a standard for
exterminating the faith and people that were called Turkish.

The Christian religion, which was taught year in and
year out, was transformed more or less into a boring subject, depending on the instructor and the material to be
learned.

In the third grade we had to learn, quite thoroughly, the
entire church service, as though we were being trained for
the priesthood. Father Jagoš Simonović obdurately required that every student, without fail, should learn the
service, if not exactly like the priests themselves, then at
least enough to be able to sing the responses. Gaunt and
severe, zealous and touchy, he was very jealous of his
dignity and was proud, like most priests of his generation,
of his fancy rhetorical style and argumentation. He could
not stand having his subject only superficially learned; on
the other hand, neither did he want to give failing grades
in a subject such as catechism. This led to some stormy
scenes, in which he was hardly subdued though he had to
appear mild in order to preserve at least a modicum of
harmony with the teaching he professed. This incited us
not only to ignore his subject, but to exasperate him. The

material he taught was such, even without his intolerant efforts, that it inspired thoughts concerning the super-fluity of church ceremonial and prayers more often than a justification of faith.

Nevertheless, we had begun to believe. Some more, others less; some through the worship, others because they were against it. We already had from our folk traditions a belief in good and evil and in their struggle in the world and within man. The new Christian teaching did not conflict with, but reinforced, by its teachings concerning suffer-ing and mercy and an idealized God, the conviction that all worship was superfluous.

Once, before the entire class, I engaged in a discussion with Simonović on this very point—that it was not impor-tant whether or not one goes to church and prays, but whether one believed. I stubbornly stuck by my opinion. The rest expressed their agreement. Because I knew his subject but poorly, I drove Simonović out of patience and made him angry. I felt that he hated me. This was not true, of course. But I caught him in his weak point—im-patience. I wanted to take revenge on him. The revenge came spontaneously, all by itself.

A good student in everything else, I had a failing grade in his subject. It was generally known that in this subject no one ever received a failing grade on a second try at the examination. However, because I had publicly belittled the subject, it was obvious that the priest was going to flunk me unless I knew the material. In the course of three or four days I learned everything and applied for re-ex-amination. The priest was quickly convinced that I knew the subject. He was obviously pleased, but he was offended because I did not look at him while answering. I found him distasteful, with all his feeble advice, imagined elo-quence, and nervous fussiness, and so I made up my mind not to look at him. He caught me by the chin, lifted my head, but I would not look at him. I even shut my eyes. He began to shout, to scold, but all in vain. He even tweaked my ear. If God is inevitable, why was this servant of His so impatient and overbearing?

Simonović was, actually, only a typical Montenegrin priest—true, educated and sound in dogma, but accustomed to having the younger generation and his inferiors submit to his will and his conceptions. Both then and later, he was very active in the political arena. After the dictatorship, he belonged to the party in power, and was even one of its local leaders. During the war, however, to everyone's amazement, he made common cause with the Communists and, though he had bad lungs and was sickly, he endured to the end through all the difficulties. True, his nationalist teaching, not his religious, proclaimed that one must always and unconditionally fight against the enemy forces that are occupying the homeland, and this he carried out. Just as Simonović consistently defended his religion, his vestments, prayers, and incense burners, so in war, he consistently defended his nationalist beliefs.

We grew, and so did the religious problem, for everyone in a different way, but ever more serious and complex for all.

If God exists, why are men so cruel to one another, so selfish and wicked? If God does not exist, is not all then permitted?

It would have been senseless to pose such questions to Simonović, even if they had been sufficiently formed in us. Whether unable to reply or, by chance, indisposed to discuss them, he simply silenced us.

Archpriest Bojović of Berane, whom we had from the seventh year of high school, was just the person for such discussions, not only because he taught us Christian dogma and ethics, but even more because of his personality. Bojović was extremely reasonable and well read, and eloquent as well. Had that been a time for great church preachers and the country receptive, his fame as an orator would surely have gone far beyond the borders of his diocese. In speaking, he sought, and found, vivid and memorable phrases. His speech flowed like a clear brook, or like honey, as the folk saying goes. In addition, he had a pale and worn face and was known as a completely chaste man and as one who never intervened in local political squabbles and intrigues. Confidence and warmth were inspired by his fine features, seemingly chiseled by inner suffering, and his small trimmed beard.

Rare was the one among us who doubted the existence of a power that exists in all things as a law. In other words, we believed in God. More important for us were the proofs of the existence of that power, and for these we searched everywhere. Archpriest Bojović was not angry or even amazed when we demanded proofs of God's existence. He apparently regarded it as quite natural, especially from young people. He answered calmly and reasonably, his proofs, in the main, similar to those of Dostoevsky. Mercy, which inevitably exists in man, is proof of God's existence. The argument was very moot, but convincing—for those who wished to believe. Man himself feels what he can and what he cannot do; there exist within him certain moral restraints. That is God. The proof of God's existence must first be sought in man, in his inner ethical categories. The very existence of these categories proves that something inscrutable and foreordained regulates man's destiny. These and similar proofs offered nothing new; their strength lay more in the way they were presented—in a beautiful and patient and, if one may say so, noble exposition.

One could expect, with reason, that Archpriest Bojović would be troubled if, contrary to his concluding God's existence on the basis of man's nature, someone advanced the opposite contention that natural laws, which regulate everything, also control human destiny. He did not deny those laws nor their influence in human affairs. He simply observed that the very existence of these laws, the very

fact that man is not able to alter them, demonstrates the
subordination of his life to forces that are independent of
him and that some call divine, others call natural. Man is
mighty where he can decide, that is, in his own destiny.
That, too, is proof of the divine force within him, and
hence of the existence of God. To the degree that he orders
his own life, man acts as if he were the Deity himself. True,
he does this according to higher laws. Yet he likewise
decides what his life will be like, that is, his life among
men. Again everything comes down to man himself—in
him lies everything, and he is the proof of both his divine
attributes and the Deity Itself.

Such explanations by the Archpriest were hardly in con-
sonance, not only with a community in which there was
precious little mercy, but also with his role as a priest—with
stole, prayers, and incense burners. Tall, gaunt, and prematurely gray, he gave the impression, as he stood at the
altar in the old monastery of St. George's Columns, of a
fresco come to life.

What connection did his rational arguments for God's
existence have with all this ceremonial, which would have
been amusing if it had not proceeded from the obscure
need of the masses for symbols and from accumulated historical tradition? Archpriest Bojović never offered any
miracle as proof of the existence of God. He would say
that men had faith without ceremonial, within themselves.
Ceremonial existed only to remind men of the divine obligations within themselves. He observed this ceremonial with
the same fervor with which he argued for God's existence.
The saints were for him men who had done good works,
preached righteousness and mercy, and that is why we remember them today, as examples.

Archpriest Bojović's explanations were completely in
harmony with the youthful disposition for justice and
mercy. He could in no way have directly influenced the
trend in favor of Communism. Yet he inspired great

thoughts and feelings concerning justice and mercy, which, in addition to other factors, of course, especially, insofar as I was concerned, the surrounding reality, led toward Communism. That was strange, for the Archpriest's arguments were designed to turn others away from Communism and every form of violence. But the desire for justice, equality, and mercy gave rise to reflection and efforts to create a world in which these would be a reality.

Later, I always felt within myself that I owed an unpaid, Communist debt to Dostoevsky and to Archpriest Bojović, a debt I did not dare acknowledge even to myself. Were not the first impulses toward Communism those arising out of a desire to put an end to the world of force and injustice and to realize a different world, one of justice, brotherhood, and love among men?

Whereas Simonović demanded that we accept what he told us as a representative of the church, Archpriest Bojović never demanded anything, not even that we believe in God. He thought that men must believe in something, in any case, and he regarded it as his task, and the task of every man, to encourage others to be persistent in their faith, that is, in mercy and justice.

During the civil war Simonović found himself on the side of the Communists. This was unexpected for those who did not know him well, especially in view of his previous political affiliations. Actually, firm in his nationalist convictions, he was really being true to himself. Archpriest Bojović remained apart during the war, not actively helping one side or the other, and certainly must have wavered, like the rest. In keeping neutral, he, too, was being loyal to his own principles.

To be sure, life itself had a stronger influence than any preachers on those who were disposed to oppose brute force.

Small-town spite did not respect its own shibboleths.

All the churches in Montenegro are built of hard cut

stone. Most are recent, since the liberation from the Turks. They are small, like one another, but striking and pleasing in their compact and modest beauty. One day they will be eloquent gemlike monuments to the tortuous liberation and hard rise of a shepherd people. They served almost as official gathering places, but only on holidays and for special rites. Generally, the taverns were far more attractive and popular. This only increased, at least rhetorically, respect for the churches as a national heritage. Still, this respect for the church did not prevent the town from degrading all that was human in a certain girl.

There lived in Kolašin a feeble-minded, sloppy, and fat servant girl, a native of Morača. Some town cutup seduced her in the bell tower in the middle of the day. Someone noticed and, as the news sped from mouth to mouth, half the town emerged at the windows, doorways, and in the street, not to prevent the desecration of a holy place, but to enjoy themselves, to snicker and to leer. Everything seemed made to order for a scandal: the little church and bell tower were elevated on a hill overlooking the town, and the principals were not quite normal. The police were informed, and two policemen started out toward the church. In the meantime, the culprit was warned by one of his buddies and he began to run through the garden, in the sight of everyone, his trousers in his hand. The poor girl was led through town. People shouted at her, not reproaches, but crude jests. She walked calmly, even shamelessly, as though not comprehending the whole affair. They soon caught the lover, but he went along laughing, replying in kind to the jests, and praising the qualities of the seduced girl.

The whole thing was revolting—the participation of nearly half the town in something of the sort, the very idea, the deed, and the denouement, but most of all the mocking of the unfortunate girl. Must men be so devotedly fond of filth? If they are not fit to point a finger at the filthy doings of their neighbors, they could at least not take delight in them! They expressed horror at what had happened in the bell tower, but they were in fact titillated by it all. They said that the bell tower was desecrated, but in fact they were simply reveling in the scandal. Nothing else seemed important to them. Being feeble-minded, she was seduced, of course, for she, too, had been given a body and desire. The attitude of the masses was more repulsive than the crime by the pair. They clucked over the poor half-witted girl, but what does not go on in their own matrimonial bowers, what dirty tricks they play on their own neighbors, and in full possession of their senses!

Are such beings molded in the image of God?

The small-town atmosphere of Kolašin was raw in its nakedness; in Berane it was more despicable by its refined talent for invention, intrigue, and insulting, ruinous nicknames which stuck for life.

Over by the gypsy quarter there lived a certain Lutvo, a porter and handy man. Nobody paid him for his work, but they gave him alms. His mother, also, worked where she could. Lutvo was a frail youth, industrious, and prone to quarreling. He was an epileptic and suffered from frequent attacks. The worst of it was that during the time of the attack, while he was in convulsions and choking, nobody would help him. He would thus lie for a whole hour, and nobody would even pour a can of water over him, let alone pick him up and get him out of the way. The majority stared petrified at his convulsions, but there were those who made jokes: "Nothing ever happens to these fellows. . . . If it really hurt, he wouldn't be jumping around like that. . . . He must have been cursed. . . . Sick as he is, he can eat for two. . . ." Lutvo would bleed from the gravel, and no one would prevent it. When he regained consciousness, he would sit on the ground and remain there a long time, slowly recuperating, and be ashamed. It was then that the crude jests and mockery would really begin, as though those who did it took special delight in having him hear their remarks. Completely broken, Lutvo would gather his strength and silently, quickly hobbling, he would flee from everyone's sight.

Lutvo was mistrustful and avoided entering into conversations, always afraid of trickery or ridicule. I wanted to act differently toward him than his mockers. Slowly he gained confidence in me, and his mother began to be pleased whenever I came. But an end was put to my Samaritanism by the very one for whom it was intended.

Near Lutvo there lived a certain Jelić, a normal-school student, already grown and a very strong, swarthy fellow. Thick-necked, with wide shoulders and powerful muscles, he terrorized the whole school and neighborhood as a bully.

I came across Jelić as he was tussling with Lutvo. Poor Lutvo was not even close to being Jelić's equal. Nevertheless, provoked and mad with anger, Lutvo not only agreed, but himself challenged Jelić to a wrestling bout. Jelić gripped him, lifted him, and then roughly smacked him against the ground. Jelić delighted in showing off his strength with this puny invalid. I dared to intervene: "It isn't fair, you so strong . . ." Jelić scornfully snapped at me to be on my way. He was obviously reveling in the triumph of his strength. However, Lutvo, evidently provoked by my assertion that he was weak, lunged at me. He suddenly took out all his accumulated rage against me and, livid, turned on me. I ran away as he threw stones after me.

I met him after that. He did not attack me. It seemed to me that he was sorry, but I could also sense that if I approached him, he would jump at me again. Life and people had so embittered him by that time that it was quite the same to him to attack somebody he did not hate.

There was another handy man who was a Moslem. He carried sacks, split wood, scrubbed floors, brought white-

wash and clay, and tended gardens. Hardly anyone knew
his real name. Everybody called him Pometina, which
means the afterbirth of a cow. Nowhere can one find such
horrible and exact nicknames as in a small town. Such
names do not show the slightest mercy, and they become
final, like an executed death sentence. Most shattering of
all was the fact that this man called himself by that name
and made peace with it. Tall and thin and incredibly unas-
sertive and good, he dragged himself through life. He had
many children, whom he fed with difficulty. Yet he was
a tireless worker, though a little slow, hungry and worn
out. Never have I beheld a deeper human sorrow and a
more final awareness of hopeless poverty and misery than
in his eyes. Once he declared, "Here I am, nothing but an
afterbirth, barely clinging to life, and yet so many gaping
mouths at home to feed, waiting for miserable me to drop
something in their beaks."

What, after all, is the aim in life of such a man? To
make a living in order to give sustenance to others, to his
wife and children? And they will do the same when they
grow up, and will be the same wretched beggars clinging
to life, unhappy because they have so long to wait for the
end.

There were some apprentices in the town. They all
dreamed of being masters, they dreamed up tricks to play
on the peasants, to the great merriment of the town, and
they despised the really hopeless poor. They were the
main culprits in inventing those murderous nicknames.

There lived in Berane a certain teacher in a school for
girls, an old maid, and in every way ungainly—too tall,
limbs too thick, and a long head. They gave her the name of
Ićindija. This is what Moslems call the muezzin's evening
call to prayer, a very prolonged chant. The girl was made
quite unhappy by this, but the more she protested against
the nickname, the more stubbornly it clung to her. Some
people would even, out of ignorance, address her as Miss

Ićindija. The town acted the more mercilessly because the girl possessed a nature of inexhaustible patience. That nickname made her whole life unhappy and lost. What young man would be seen paying court to a girl whose name was the joke of the whole town? They would stick him with a nickname, too. Who would marry such a girl?

Such nicknames, inventions, and jests are given supposedly out of fun and without malice. Those who are on the giving end do not mean anything bad by it. But evil and vileness taint both those who suffer such jests and those who make them. Both sink into vileness and bitterness.

If religion is unable to better human relations, nor are so many wars and rebellions able to do so, is it perhaps because nothing at all can be done? Still, men do work at this and succeed just the same, even if only a little. What will be the force that will bring about the great transformation?

Young people—each in his own way—pose questions and seek answers. It would appear that the solution lies in this constant search. But everyone wants to find nothing less than the final solution in his own time—especially those among young people who are dissatisfied with the state of affairs they find, and who are sufficiently strong and serious to look social reality straight in the eye.

There is nothing finer and more delicate in this land than a sister's love. Montenegrin mothers love like mothers everywhere, except that they are perhaps more easily reconciled to the death of their children, especially in war or disaster. Mothers are mothers, even in Montenegro. But Montenegrin women love their brothers—even their cousins, if they have no brothers—with a love that combines a feminine feeling at its purest and subtlest with a primeval determination to preserve the breed from which one has sprung. That tongue of stone and fire which knows no words of endearment becomes transformed in the mouth of a sister into an incredible softness and cooing. A sister is not something greater than a mother, or less—but different. She has a more direct and irrational warmth in her love for her brother. A sister will quarrel with her brother, but she will never break with him. She does not share with him in the property. The family has no obligations toward her, or she toward it. She simply gives and accepts love and goodness. Her love is purer than any other except the love of blood brotherhood. Her love rests on tradition, on feeling, on an inherited gift. It does not falter, but grows with the years, with death and calamity. It is a constant and limitless sacrifice and joy which always finds it possible to sacrifice still more. Such a love my older brother and I hardly felt in our childhood and youth from our cousins. Our sisters were much younger than we, and only after we were grown up did we feel the full yearning force of a sister's love.

Love among brothers is different, though it need not be any weaker. Love between a brother and a sister never has great obstacles. Love between brothers is rarely without difficulties. Brothers quarrel as children, as youths, and as mature men, and over everything—over play, over seniority, and over property. Brothers can also come to hate one another profoundly. Yet even then there remains between them an unquenchable spark of love.

In Berane, our quarrels were as frequent as they used to be in childhood, though we ourselves had changed since then. My older brother was already a young man, and I was no longer a boy. Wrangling and tussling had been an integral part of our games among peasant children. Now, however, he took my resistance for an insult, as I did his thwacks and beatings. Now neither insults nor blows could be easily forgotten.

I was becoming more independent, and, through reading, my ideas were changing quickly. The changes in him, however, were much more profound. He filled out rapidly, began shaving like a man, wore a tie, and grew an enormous shock of hair. His limbs became strongly developed, so that every muscle showed. From a puny lad he had turned into a strong youth. His sensitive nature became more tolerant, though, patient until his boiling point was reached, interspersed with sudden bursts of rage. It became obvious that he was interested in women. He did not hide it; he was open in everything.

It was not easy to become a young man in Berane, especially for a wild and free personality like my brother. Through hurt pride he clashed first with his home-room teacher, and then with all the other instructors. Other uninhibited and impetuous youths, especially those from the mountains, had the same trouble, and often some rude outburst led to expulsion from school or the abandonment of an education. Hatred and conflict developed easily, especially between the older students and some of the instructors.

My brother felt cramped in the school, in the tiny room in which he was forced to tether his growing strength, in the half-peasant clothing in which he felt degraded, for he yearned to be dignified and distinguished. He was bound by our father's thrift, my own striving to be his equal, and by the frustrations of a small town which was gray and dull with no women but worn widows and waitresses. He felt that he was destined for something bigger and better, that he could achieve nothing there.

The last and only major, unforgettable clash took place between us at Christmastide in our parents' house. We quarreled over some trifle, and he turned on me in a house full of guests and family. He could not endure being crossed by a youngster, and I could not endure being struck. I grabbed a knife from the table. The stab in his thigh was deep and wide; the blood spurted across the room and across all my thoughts and perceptions. It happened almost accidentally, probably like most crimes. Having taken up the knife without any intention of striking, I could no longer lay it down for shame, while he could not give way for similar reasons.

Perhaps this was precisely the occasion, though a horrible one, for a turnabout in our relations. After that he never turned on me. Apparently he understood, all of a sudden, that I was sufficiently mature so that beatings could solve no disputes. From then on there was always a bit of sadness along with loud joviality and fun whenever we would meet and an irrepressible yearning never to part. Yet we had to live apart. We were separated finally by his death in October 1941.

Almost the identical story was enacted between my younger brother and myself. I was in the seventh year of high school in the spring of 1928. My younger brother,

Milivoje—Minjo—was going to school with me in Berane when a conflict arose between us. I began to beat him, and as he defended himself, I could not stop. I struck him hard, with all my strength, pummeling his back with my fists and slapping him. Finally he gave in, and the blows still fell. Then he began to cry, as grown men cry when they have been humiliated and yet are powerless. I remembered that it had been the same with me when my older brother beat me when I was fifteen. Now it was happening again, only I was the older. We never clashed after that. That beating, actually the last, could not be forgotten. But he forgot. He was executed in Jajce in 1942, after prolonged torture in the police station; the only words he would permit his tormentors to enter into the record were: "I wish to die an honest man and therefore will not betray."

The love among us brothers expressed itself first in mutual confessions, new discoveries, and endless discussions.

In the summer of 1926 my older brother and I were particularly close. We lived different lives; he rushed to meetings in towns and trysts in the hills, while I read and went fishing for trout. Yet it was a life together, for we confided to one another experiences we did not consider complete until they were shared.

We would sleep togther, in the shack by the sheeppen, in the dew and the moonlight. At night, as soon as we would lie down, he would slip away to the village, to some village belle. I would fall asleep immediately. He would return at dawn, wake me up, and tell me all, without keeping back anything. Then we would fall asleep again. Once, a woman with whom he slept the night secretly slipped two hard-boiled eggs in his pocket. He laughed at that as though it were something childish and simple-minded; she had wanted to thank him in some way for his affection. In Montenegro a life of intimacy is not shared with others.

In this respect Montenegrins are all austere and completely reserved. It was through my brother that I came to know of such a life and of the ways of men and women. Reserve was only something on the surface, for the sake of the community and self-discipline; underneath that, people loved and suffered just as anywhere else. So it was with my brother.

My brother's difficulties in Berane grew worse. He was failed and forced to repeat a year. His resentment against the instructors, the school, and the community constantly increased. Without asking Father, in the middle of winter in 1926 he withdrew from the school and, with only a bit of money, went out into the world. He did not stop until he reached Belgrade. He finished that school year and normal school in Belgrade as one of the best students. The new environment, in which he felt freer, had a decisive influence upon him.

He became more elegant in his style of living and more selective in his amours. He no longer wished to live in our village, in the wilds, but moved to our new property near the town of Bijelo Polje. He dressed nattily. Pale, with long and thick eyebrows and distinctive features under a wide black hat, my brother was a handsome man, and women and girls both fought over him.

In the fall of 1927 I became very sick with typhus. My brother was in his last year of school in Belgrade. His letters, which I read only after my recovery, revealed to me how stirringly profound his concern was. The letter he wrote on hearing of my recovery betrayed a sorrow that was all the greater because it expressed a heroic joy.

Other elements were also involved in our love—political, philosophical, and spiritual—making it all the firmer and more alive.

On his return from school in Belgrade, in the summer of 1928, he told me of the demonstrations at the Russian Tsar

Hotel in Belgrade. They had begun as anti-Italian, against the Nettuno Convention,* and then the Communists and other opponents of the regime became involved. An attack by the gendarmery sharpened the conflict, and the demonstrations shifted from foreign to domestic significance. His unrestrained nature and his vague political discontent (some of it on nationalist and some on social grounds) impelled my brother into the riots. He was crushed by a mob that had been chased by the firing into the hotel cellar. They carried him out unconscious, and there he recovered. A gendarme declared, "Let me knock the fellow over the head with a rifle butt to teach the others a lesson," but a hospital attendant intervened and got my brother safely into a car.

This was the first great political experience we shared. I participated in it through my love for him. I was moved against the existing regime particularly because the gendarme wanted to bash in my brother's head for no good reason, and during a patriotic demonstration at that.

Then came another event, which was to be far more significant for us—the murder of Stjepan Radić and his colleagues in the National Parliament in Belgrade in June 1928.†

The murderer was a Vasojević, Puniša Račić, a native of the village of Slatine, near Andrijevica. At election time he hung around Berane, especially on market days. Stocky and swarthy, with a trimmed mustache and a gold chain across his vest, he might have been a cattle merchant or a coffeehouse owner or a member of Parliament. Actually

---

* In 1924 a five-year "peace" pact between Italy and Yugoslavia was concluded, and in 1925 the so-called Nettuno Convention was drawn, designed to implement the pact in economic and cultural ways. Croat and Slovene deputies, still objecting to the annexation of Fiume by Italy after World War I, blocked its ratification for three years. The Nettuno Convention was ratified by Yugoslavia in 1928, but by then it was meaningless and served only to set off anti-Italian demonstrations in Belgrade which were suppressed by King Alexander.

† See note on page 230.

he was a *komitadji* * and a political assassin, a brawler, and
quick to draw a gun. I saw him quarreling once in front
of the coffeehouse with some political opponents. Livid,
he bellowed, "Who's a liar? ——his father!" Now his shots
rang out in Parliament, mortally wounding an already frail
and unripe freedom.

At that time I was in the sixth year of high school and
already a reader of newspapers. No other student in the
high school was, and the instructors did not look kindly
upon it. The newspaper *Politika* published the stenograph-
ic record of Parliamentary proceedings, which I constantly read. Stjepan Radić's influence had been growing
unimpeded among the common people, because of his skill
and perseverance as a critic, and even more because of the
dissatisfaction of the Montenegrin masses and most zealous
adherents of the unification. The actual power was, in fact,
already in the hands of the police, while the politicians
simply vied with one another in hawking promises from
election to election. Graft, the plundering of state property,
scandals of every kind, even trafficking in national interests,
rumors of death by torture, the high living of a handful
of rich men, and debauchery in high places—exaggerated
reports of all this came to our region and into remote villages and small towns. To ordinary people, who were not
infected with partisan passions, Radić, though only the
leader of the Croatian peasantry, appeared to be the vigi-
lant conscience of the entire country.

It was obvious that as Račić drew his pistol to shoot at
the Radić deputy Pernar, supposedly because of an insult,
he aimed in reality at Radić, and at Parliament. The shooting in Parliament turned Radić into a martyr for liberty
in the eyes of all honest people. But it also buried Parliament
morally.

Both my brother and I, I remember, were struck by
Radić's murder. We sensed that something more important

* A *komitadji* is a rebel, a military irregular or guerrilla.

and ominous than the assassination of a political leader was involved. We had no great illusions about the ability of the King to help make things right. As a matter of fact, it was rumored that Račić had ties with the court. The King's visit to the wounded Radić in the hospital seemed to us not genuine. Why does not the King do something great? He alone is able to. But he did nothing great. He, the King, was at the head of the plot against Parliament.

My brother and I were bound not only by a deep brotherly love, but by a common resistance to the given state of affairs and conditions. Such a relationship could only grow stronger through the long years, regardless of whether we were together or not. We developed politically at a different pace, but in the same direction. This strengthened our love. There was nothing that divided us; everything bound us together. Without this love for him I hardly would have known what real love was, the kind that nothing can destroy or throw into oblivion. Perhaps his death before a firing squad was made easier if he had time to think of my love.

Our love became the core around which the love among the other children was entwined. Even the relationship between my father and mother settled down to become a beautiful mutual devotion. Ours would have been a happy family had not imprisonment, persecution, and murder torn first one and then another member from the whole. Not only did our love unite the rest of the children, but our views attracted them. This was first the case with our younger brother, Milivoje. Then the other children followed suit. Of the seven children in our family, time turned them, one by one, into seven Communists. Such was the fate and the path of other Montenegrin families, and not only Montenegrin. Each enticed the other through love and like views into struggle and death.

No matter who came into contact with the two of us, or for that matter with our family, regardless of how close he might be to one of us, he would immediately feel himself outside of a closed circle which it was impossible to penetrate. The jokes of this family circle, their observations, forebodings, ideas, play were incomprehensible and repugnant to others. My wife, Mitra, always felt isolated and unhappy among us, though we all loved her and were attentive to her.

Apparently it is always like this wherever there is much mutual love and people shut themselves up in their own world, one which is for them perhaps the best of all worlds, but which others find inscrutable and unbearable.

Now there are no more brothers, nor that happy circle, and others need not feel excluded any longer. One form of life is mowed down, and another has not yet sprouted. Silence remains.

The comrades are all gone, too. Some have died, some have been killed, most have disappeared into the humdrum of life; from some I have separated long ago, from others just recently; I have taken a path that was not theirs.

So it is surely with every man. But each man bears within himself, unaware, the imprint of the lives of those whom he has loved. And also the painful brand of his enemies.

One had to live in the city, in Kolašin, to realize that even among city people, that is, among city children, there were fine and good souls.

Milutin, Mićo Zečević, had as a child all the characteristics he exhibited later as an adult. A scrawny and goggle-eyed kid, with wide uneven teeth that gave his smile an unusual charm, he was sick with angry righteousness. Mićo's family was very poor. His mother kept a little coffeehouse in their home, and his father, who had lost his eyesight in an American mine, was a trial and sorrow for the whole town, let alone for his family. He was a reasonable and good man and in full possession of his strength, and therefore felt like a burden and nurtured a dejection that infected his entire family.

There was something morbid, forlorn, and shattered about that family, and particularly about Mićo. He seemed to be a spiteful child on first impression, but this was because he was so distrustful of strangers. He was unusually orderly and among the best students, at bottom proud and willful, but not a troublesome lad. His physique was frail and undeveloped, yet he was unexcelled with a slingshot. This same accuracy was evident in his thought and intuition.

Like all city children, he hated the peasant children and took part in attacks on them. They feuded with one another, the city children united against the peasants. I was recruited immediately into their militant ranks, but when I would see the peasant children, frightened and scattered, fleeing in their tattered clothes, I would remember that there were among them friends with whom I had gone from the village to high school, and I would withdraw from the fray.

Among those peasant children, Mijat Mašković and Mihajlo Četkovic were dearest to me. We were in the same grade and waited for one another on the way to school. It was quite a long distance, a whole hour's trip, and that distance brought us closer together. We were bound also by common games, fights, and the winter cold. Both boys were war orphans. Mijat was taciturn, somber, and withdrawn. Mihajlo ruddy, freckled, and aggressive. Mihajlo learned with difficulty, Mijat with great ease. One knew they were orphans and poor; their joys were few and they took offense more quickly than the rest. It was easy to bring them to tears, and hard to console them. Like Milutin, both were known for the slowness and caution with which they embarked on anything new.

Not one of my school friends, neither those in Kolašin nor those in Berane, became an active Communist when I did, immediately at the beginning of university studies. They joined only after graduation. Zečević was a doctor when he decided, with unusual seriousness and conscientiousness, to join the Communist ranks. He remained there with that same seriousness and passion until his death —the Germans hanged him near Obrenovac in 1941. Even as a boy, one could foretell that he would be a man of conscience and feeling. Mihajlo also joined the Communist ranks late, though he leaned to the Communists as a lad. His relatives were Communists, in fact the only ones in that district. He was shot in the war. With him, too, there was no sign in his childhood as to his future political affiliation, but it was obvious that he would be an honorable and prudent man. Mašković, too, became a Communist later. He was arrested in 1936, though only for a short time. He went to Spain, where he fell.

These were not my only friends in childhood. I had closer ones. But they were the most significant, and I regarded my friendship with them as of special importance. To be sure, it might have been chance that they, too, became Communists, independently of me, as I of them—each going his own way at the same time. On the other hand, the majority of my schoolmates, even those closest to me, were either hostile to Communism or utterly indifferent to it. In Berane especially I had many good friends among them.

In the spring of 1928 there came into our class at Berane, Radovan Zogović,* who was already publishing his first works. He had been in Berane before to study, and then went to Peć, where his family had moved. He was expelled from there, though not for politics, and transferred back to Berane. He did not stay long, only two or three months.

He was rather tall and extremely thin, freckled, and with sharp green eyes. His bearing was always brusque and stiff. I was never able to get to know him well; he was all thorns and seemed to have only contempt for our local poets, of whom he spoke derisively. He, also, teased the girls. Wherever he went he cut a swath as with a barber's razor. I used to ask myself what kind of man he really was inside. What were his thoughts on life, literature, on the

---

* Radovan Zogović was to become, with Djilas, a Montenegrin *enfant terrible* in the Communist literary movement before World War II in Yugoslavia. In 1948 he opposed Djilas—and, of course, Tito—on the break with Moscow, and disappeared.

books that he had read? Why could we not get together and help one another? Actually, we never did because he made a point of avoiding the circle to which I belonged, whereas I was repelled by his acrid temper. So it happens that men never meet, though they might wish to, but move away from one another like ships on the sea.

Zogović left without my ever having come up to him, nor did he ever approach me. Neither he nor I suspected at the time that our lives would later merge in a common struggle only to separate even more cruelly in 1948 when Zogović took the side of the Soviet Communist party.

I was eighteen years old when I became aware that a man and woman can be friends without the admixture of any erotic impulses. Perhaps it is precisely because these elements, though present, are suppressed and stifled that such a friendship becomes endowed with a tenderness and solicitude that is absent among men.

Dunja Vlahović lived in a nearby village. In walking to the monastery, I would frequently stop at her house, especially in the spring of 1929. We would talk till late at night, and I would later go back along the paths between the fields, through the heavy fragrance of wheat and under the close stars, all of which I never would have noticed had it not been for my meetings with her, the kind of meetings in which people bare themselves to one another.

Dunja was one of those lithe and strong women whose curves and sinews showed forth with every movement. She had a gaunt face, almost that of a sufferer, and sad yellow eyes. She was measured and deliberate in everything she did; she was even stingy with her smiles. Deep down, however, she was one of those disguised passionate creatures who burn with a constant inner fire.

Before finishing school she was married, and very well for that time, to a district chief. Of course, like most police officials, he was on the Chetnik side during the civil war and perished as a result.

My friendship with her did not attain constancy and

stability, but it pulsated powerfuly while it lasted. It might have been just one of many had there not taken place a chance meeting between us after many years. We did not see one another from our student days to the summer of 1953. Then we met in Bar, where she was a petty clerk. She asked to see me. The local officials interpreted this request as her desire to seek advancement. This was not the case at all. She simply wished to see a companion of her youth.

I was one of the top government officials. The revolution had taken everything from her; it gave me everything— except what I had idealistically expected from it. She had already become reconciled to the misfortunes that had over- taken her; I had already begun to feel rising within me a new discontent and vexation. Long ago, as students, we had been close. Then we went our different ways, ways that could never meet without hatred and recrimination. Now, as though life were playing with us, we felt, each in his own way, a similar joy and sorrow and shame that everything could not have been better than it was.

We walked along the seashore without saying a word, as though by agreement about all that had happened since our student days. She did not ask for a thing, nor did she com- plain about anything. That erstwhile warmth between us was stirring out of death and oblivion. Twenty-five years had gone by, but the old gladness was there, as though nothing had happened that might break or poison this interrupted and forgotten friendship between two people who were different in every way and who had traveled different paths. And then, on the foam-flecked seashore, those scents of the fields along the Lim and the stars that one could touch above the looming mountains all seemed to come back.

More complicated than any other was my friendship with Milan Bandović. He was extremely intelligent, and even more cunning. That integrity which marks true depth and courage was lacking in him. He, too, was one of those students who worked little and achieved much. Very poor, he was aware of his considerable talents and deeply dissatisfied that he could not develop them and attain success. Others who were obviously less capable forged ahead, and this galled him and made him bitter. He barely got through school, for he did not study regularly, and then only what interested him.

We were very close for a while, but also clashed. Two or three days after the proclamation on January 6 of the dictatorship of King Alexander Karadjordjević, the two of us waged a bitter quarrel which was unforgettable because of the importance of the event. He approved of the *coup d'état*—the King had preserved the state and unity—while I was of the opposite opinion. We were still young, and our political passions were not yet sufficiently developed for us to break because of that. However, our differences made themselves felt even then.

Ours was a friendship full of contradictions and difficulties, because it was between persons who were completely different. He was the more eager to preserve our friendship, and I carried on the quarrels, feeling that he would not be offended because of them. Once we even had a fight in my little room. It was a very serious fight indeed, at night, man to man. He was stronger and older, but I was the more determined. I finally got the better of him, and then was beaten anyway. We were not friends for a long time after that. Then he came to my house and managed, with the help of my parents, to make up.

He seemed to enjoy the role of the wily politician. As a student he at first drew near to us Communists, more through friendly ties than ideology. He was a Democrat in Davidović's party, only to turn suddenly to Stojadinović's Yugoslav Radical Union.* My brother and I and other leftist youths in our region regarded this as opportunism

---

* After King Alexander's death, Milan Stojadinović became prime minister and foreign minister (1935–1939) and oriented Yugoslavia's foreign policy toward the Axis powers.

and betrayal, and broke off all relations with him. Nor did we like it when, in 1936, after I returned from jail, he asked the district chief whether he could visit me. We both resented and laughed at him.

During the war he was a district Chetnik commander and, what is more important, the organizer of Draža Mihailović's youth movement. The capitulation of Italy found him in that country, and he returned to Yugoslavia from there. It is extremely strange, in view of his caution and agility, that he did not anticipate the danger that lay in store for him if he fell into Partisan hands. Perhaps he thought he would reach Chetnik territory. Or perhaps he thought that his participation in the Chetnik movement was not such as to cause severe punishment by the Partisans. Most interesting of all, he traveled with a group that was on its way to report to the high command. Some fighters recognized him. He was arrested and, instead of reaching Montenegro, he was placed under investigation. This was at the beginning of 1944.

The very first bits of information about him and his statements proved to be very interesting indeed to the Partisan high command. The English and Russian missions were already at that time with the Partisans, and not only the liquidation of Draža Mihailović's movement, but also the exposure of the role of the *émigré* government loomed ever more imminent.* Bandović was a man who knew

---

* After the Germans and Italians invaded Yugoslavia in 1941, Draža Mihailović's Chetniks were the only organized guerrilla force resisting the Axis occupation troops. Both Great Britain and the United States furnished supplies to the Chetniks, who were recognized by the *émigré* government of King Peter II in London as the Yugoslav national army. In 1942 Tito's "Army of National Liberation" was organized, and by 1943 the Partisans of Tito and the Chetniks of Mihailović were warring against each other. In 1943, too, the British shifted their military support to the Partisans (without, however, disavowing King Peter's government). By 1944 the United States was active in sending aid to Tito. The Soviet Union supported the Partisans from the start but more by promises than by supplies.

much, especially about Draža's collaboration with the Italians. Tito got the idea that it would be well to keep Bandović as a valuable witness for some future trial. It was convenient that I knew him well from before, and so it was decided that I should talk with him. Ranković and I outlined the details of that conversation.

After so many years of fighting on opposite sides, we met in a jail cell, in opposite roles. The talks lasted long, throughout two nights. Bandović recalled our friendship—to move me, or perhaps he himself, being in prison, was moved. I found both his emotion and my role unpleasant. Why this underhanded game between men who were enemies? Between us lay a surging sea of blood and hatred. It was not difficult at all to convince him that he ought to help us; what was difficult was to do so, supposedly as a friend, without offering him anything definite and without promising anything. This was all the more difficult for me because nothing had been decided in his case. Nevertheless, I hinted to him—in the usual way—that his head was saved. However, just as it was clear to him that his statements might be the price for his head, so it was clear to me that it was not I, a former friend, that had softened him up and made him more amenable to the demands of the inquisitor, but the fact of prison and an extremely precarious fate. Our former friendship was only an excuse in a game that both he and I saw through, but which both of us, for various reasons, wished to win. We were both on our guard not to talk, at least not openly, about all of this—I about political interests, he about saving his head. His bright, wise, black eyes comprehended what it was I was trying to conceal, though he pretended that he did not see through me. Knowing that he was no hero and not firmly grounded ideologically, I knew just how far he was able to resist. We knew each other well and were both too penetrating not to understand one another without much explanation. For this reason we talked aimlessly, avoiding memories that would recall the warmth of our former comradeship, and even regretting and being ashamed that it ever existed.

After that the investigation of Bandović offered no problems, especially since it did not make a special point of dwelling on his personal guilt. Bandović's statements were deemed quite valuable, and could be considered so until Draža Mihailović later made his own insignificant confession.* Anyway, Bandović got away, but with no thanks either to his own skill or to our gratitude—in which he disbelieved in any case—but to the confusion that overtook the guard, who dispersed at the beginning of the German descent on Drvar on May 25, 1944. The Germans freed Bandović and he managed finally to emigrate.

It was misery that led many to take up the Communist path. With Bandović it was different. His inability to make good in life, to extricate himself from peasant apathy, primitiveness, and mire, caused him to set out on a path he never would have taken otherwise. He was born to be a good professor or, in happier circumstances, a wise, capable, and canny statesman, and instead he became one of the leaders of a movement that was never really formed, again in a primitive and backward peasant environment, and finally to eke out his days far away from his land and the dreams of his youth.

And so it went. The majority of my friends never finished their schooling, but as petty clerks sank into the raw peasant masses from which they had tried to emerge, or else they were lost in the steep and dangerous bypaths of this wind-swept terrain and treacherous clime.

The deepest and most touching friendship of my youth was with Labud Labudović. It contained nothing but a comradely love which was simultaneously both passionate and rational. He was the best and most modest student in

---

* Captured and brought to trial in 1946, Mihailović confessed his opposition to the "People's Front," the Partisan movement that formed the postwar government.

the class, and perhaps in the entire school. School was not easy for him; he achieved success through stubborn effort and unbreakable will. He was the son of a schoolteacher and came of a retiring and respectable family. They lived in town, largely off the father's modest salary, in an out-of-the-way section but in a house that nestled in garden greenery. Everything about Labud and within him bespoke quiet and solitude. None of our crowd liked him. He, on the other hand, did not go out of his way to gain anyone's favor; he was too proud to beg and too discriminating to display his feelings. He was a rather handsome, ruddy, clean-cut, and neat lad, who paid special care to his fingernails, something none of us did. As an outstanding pupil he was the favorite of the instructors, and a good son. He had a single basic trait for which my crowd disliked him: he was too orderly, in an environment that was uncouth and unkempt. He appeared to be the incarnation of a confining and soulless pedantry. He was, in fact, orderly and pedantic out of an inner congruity, sensitivity, and purity, and not out of vanity or pettiness. Yet he did overdo it.

We became friends through a feeling of rivalry for which he was more to blame than I. I was better than he was in the Serbian language, especially in written assignments. He might have been rather cool toward me because of this, and the class egged him on, for all were jealous of him, not only because he was a better pupil, but also because of the rather favored circumstances under which he studied. Besides, they could not stand him because they interpreted his solitude and neatness as the arrogance and insensibility of a town dandy and teacher's pet.

The clash, or, rather, our friendship, began directly as the result of a sharp discussion in a student literary club. I did not like to go to its meetings; they seemed too schoolish and monotonous to me. However, when it was Labud's turn to read his works, my friends persuaded me to go and to squelch him. Perhaps I was not his better in the discussion, but he was so struck by my criticism that he could not say anything but instead began to stutter while his eyes blazed with anger. As it was, his voice was weak and hoarse because there was something the matter with his throat. Now he was reduced to a painful stammering, a vain effort to wring out a word, let alone a consecutive sentence. He left the classroom, and I, agitated, followed him. I felt guilty for attacking him, and even more for letting my friends goad me into doing it. He noticed that I wished to speak to him, but he did not choose to stop. He was already turning into his yard when I rushed up and called out his strange and lovely name, a name that suited him so well. Labud means swan. Thus began our friendship, in the sixth year, and it lasted up to the beginning of the eighth year.

We would spend whole afternoons, and on Sundays and holidays the whole day, walking around town, engaged in long conversations about books, about our plans, not hiding a single thought or a single emotion. He would wait for me in the morning to join me in those two or three hundred yards to school. In the evening we would part sadly, each going a part of the way with the other countless times. It was a friendship that stirred and intoxicated. It was then that I came to know how subtle and complicated his soul had become in that hard shell of proud isolation.

We, too, finally had a quarrel, over a trifle, completely unimportant. But we could not set it right again. He was in love with Inge, the daughter of Mrs. Ugrić. The whole school was in love with Inge. I, too. Everybody fell in love with the pretty and enticing Inge easily and transiently. But Labud's love was until death. He understood everything slowly and seriously, deliberately. However, in his love for Inge all of these qualities were hindrances rather than advantages. He hid his love, and it ravaged him all the more and drove him into seclusion, into somber melancholy and wild efforts to free himself which only crushed him all the more. He loved Inge even later, as a university student, after her marriage. All this had an importance for him that nobody, except perhaps myself, could even suspect. This was the love of a proud soul which could never admit that it had been conquered. That love crushed and consumed him.

He did not speak even to me about this love of his. But I knew about it—by the blush that covered him at the mention of her name, by the way in which all paths seemed to lead him to her. Labud carved into the desk in front of him an emblem nobody understood. However, I puzzled it out, knowing of his love: it was Inge's name and his—the letters being intertwined. Once they asked him what it was. He was evasive. I said that I knew, but that I would not tell. He claimed that I did not know. An argument ensued, and I told. He flared up and declared that I had made it all up just to humiliate and slander him, but I knew that I was telling the truth.

The quarrel was unimportant, even though the reason for it was a grave one. Nevertheless, after that incident, neither of us managed to effect a reconciliation. The conflict grew with each day as it devoured both him and me.

The paths we had walked together and the springs we had discovered vanished without a trace; the hills we had climbed sank into darkness without him. Everything brought him to mind, with a certain bitterness which goaded me into senseless rejection.

We did not speak to one another for nearly the whole year, despite the emptiness and unhappiness that we endured. It was not a question of who would give in first. Either of us would have been willing. Something else was involved: each of us was, in fact, angry with himself for permitting such a great and wonderful friendship to become troubled, even for a moment. We always spoke well of one another before others and even tried to help one another. Labud went so far in this that he even slipped me answers during written exercises, which I refused haughtily

but not rudely. I did similar things for him. These tacit
favors, however, did not weaken, but merely reinforced,
our quarrel. We would seek one another out, for months,
and then turn our backs on meeting. Both of us suffered.
Yet, as though we were cursed, we could not make up.

The entire class knew of our break, even though all had
forgotten its cause. They tried to reconcile us, but nobody
could. Professor Zečević tried, also. Devilish as he was, he
hit on a clever approach: you fools are fighting over her,
and she is deceiving both of you. We knew, however, that
she was not the cause of our quarrel, that it was not jealousy
that divided us, but the pain of self-punishment for allow-
ing ourselves to hurt the other.

After graduation, when it was quite apparent that we
were going to take different paths and that we would not
see each other so frequently, the two of us shook hands.
But our relationship was already dead. It would have been
better if that handshake had never taken place, for then we
would not have realized the bitter truth—that a great friendship had been killed.

Labud died in the course of his university studies of
tuberculosis of the throat. He was not sick even three
months; he was simply mowed down. I was already known
at the time as a Communist and a young author. Labud, too,
tended toward Communism. He was too serious and deliberate to rush into anything, yet he was also too intelligent not to feel the spirit of the times. He was certainly
going in that direction. He was not a man of passion, swift
decisions, and brain storms. That is why with him love, like
anything else, was so deep and lasting.

He was buried on a hot summer day. To cover the odor,
they sprinkled eau de cologne and covered him with
flowers. His illness had sucked him dry; now there were
only those thick eyebrows, which had been knit together
with anxiety ever since he had entered upon youth, love,
and the unknown. So he went, without our coming to a

final undertsanding. This made the sorrow he left behind him all the greater.

Apparently the saying is a truthful one that mountain will meet mountain, but man with man—never. Yet man cannot do without man. The most collective of all creatures, man is also the loneliest.

Surely like most men, I, too, experienced three stages in love: in childhood, adolescence, and youth.

These stages were all, as with most men, quite usual—sincere and painful and profound, and all ended unhappily. Not one was capable of keeping me from those paths that I considered mine. There was something stronger than love or even death that determined where I was to go. So it is with everyone else—everyone has his own inescapable path. With some, however, these paths may be loves.

It probably would have been another girl had it not been for the fact that Dobrica was my landlady's daughter and that she acted toward me like a sister. I was thirteen, and she was two years older. She was not pretty, and this was quite obvious to me. I would have been both happier and unhappier had she been a beauty. Rather short and plump, with freckles on her hands and face, she had pale blue bulging eyes and stringy washed-out hair. When she grew pensive or stared aimlessly, one could just barely tell that she was slightly cross-eyed. In fact, this defect was the nicest thing about her. She was already at an age when girls begin to fill out, and she, who had been skinny until then, suddenly began to blossom.

We went to school together, walked back together, ate together, and slept in the same room—she on the bed, and I on the floor. So it was with parties and everything else. Being a good-natured and warmhearted girl, and never having had a brother, which in Montenegro, and even here in the city, was regarded as a great misfortune, she liked me sincerely and deeply. I liked her, too.

This love was different from the family kind, and it began immediately, in the fall, while we were picking plums and sitting side by side on the wall beside the house on sunny afternoons, studying and chatting. It was as though that sun tied us together imperceptibly with its gentle rays. Not even at Christmas did I feel like going home; without knowing why myself, I wanted to be constantly with Dobrica. The winter was a snowy one. Driven indoors, we were together all the time. At carnival time we went to see the mummers. We went gaily along a wide path cut in the snow. Some peasant's horse, which was hitched to a sleigh, was frightened by the masks, bolted, and came galloping at us. I noticed him too late to jump up on the high and hardened snowbank. The horse struck me at full gallop with his chest and the sleigh ran over me. For several moments I was apparently in a faint. Then I tried to get up, but collapsed. Dobrica ran to me, lifted me, and began to brush off the snow which covered me, wailing and weeping all the while. I had never seen her cry before. Her eyes bulged out even more, and that lost cross-eyed look came over her eyes. All crumpled, and weeping, I began to console her, to convince her that I was all right. It seemed to me that it was at that moment that both she and I became aware of our love.

However, that was true only for myself. She loved in me only the brother she did not have.

A good friend of mine at the time was Milosav Pulević, a lively freckled boy who was extremely prepossessing, especially since he lisped and had special trouble with his R's. We were both in the same grade, and, since he lived in the neighborhood, we were together a great deal. He had an older brother, Mihajlo, already a young man. Dobrica and I went with them frequently. Dobrica's aunt and uncle and also her grandparents lived nearby. But she also went there because of this older brother.

Milosav told me all about how the two of them, Dobrica and Mihajlo, liked one another, how she was in love with him, and he not at all with her. He had as many girls as he wanted and was not interested in her a bit. Milosav persuaded me to see that love for myself. And I did. They met in the garden and held hands.

After that I saw everything—how she acted coy in his presence, how they picked one another as partners in games, how she blushed at the mention of his name, and how she would talk me into going to see Mihajlo and Milosav, even when we had homework to do. It was then that our first quarrel took place—not just a quarrel but hatred. Dobrica was capable of flaring up, even at her own mother, and so once she clashed with me. Perhaps it was just an excuse for me to punish her with my wrath for her supposed infidelity. She immediately wanted to mend our broken relations. She became even more tender and attentive. I, however, became all the more stubborn. I was sad— humiliated, despised, rejected.

Not even a spring excursion by the school to the Morača Monastery restored my good humor. There I found new vistas—the high rocky peaks and bluish dales, the clear river and Prior's Bridge, a stone ledge that linked cliff with cliff, the grassy meadows, the moss-covered monastery perched over a waterfall that splashed gaily into the Morača. The monastery and everything in it, as well as the stories about it, was the living breath of a distant past, but a past without people. Perhaps because of these legends and its age, the little monastery seemed to loom large and gray with a filmy radiance, like a star among the mountain peaks, as the folk song said. Yet it was also full of melancholy. The fragrance and shades of the past and the gaiety of the group reminded me of my unhappy loneliness in love.

Books and painful solitude, and dreams, constant dreams, forced themselves upon me. The poems, which I had just begun to write, were all sad, and the sun set in them as though it would never rise again. Every line throbbed with sorrow over her, Dobrica, a sorrow all the more profound because it could not be shared with anyone, for to read anyone one's own verses, love poems at that, meant to bare one's purest essence, that which one does not reveal even to oneself.

Another love only aroused mine all the more. In fact, that love inflamed the entire school and stirred up the little town, which was set in its ways. This was the love between Ljubica, Dobrica's girl friend, and a certain Mučalica, whose wit and radiant good looks made him the most popular student in school. She was a black-haired Juno with a long jaw, stubborn and determined in everything, but also very devilish. He, on the contrary, was as nimble and swift as a deer. She was a city girl, and he from the country. Both were good students. His natural keenness and agility were complemented by her sensuous and listless languor. So also in physique did they complement each other—he tall, she short; he auburn-haired, almost blond, she a brunette. Countering her openness, he was as shy as a girl.

That there were secret loves among the students, all the teachers surely knew. But that any of these children dared to bring their love into the open, such brazenness never even entered their heads. The pair's letters were intercepted, their trysts were witnessed, they were reported, and finally brought before the principal. Usually in such cases the students would admit half, cry hard, their parents would take steps and the whole matter would rest. Mučalica, however, and Ljubica, too, apparently under his influence, did otherwise. He acted just as one might expect of the school hero: he admitted everything from the start and declared that he loved Ljubica and did not intend to repudiate that love.

No advice or threats did any good. Now the teachers were at a loss, and the two became the heroes of the entire school and of most of the town, which was always eager for anything that smelled of scandal. But there was no scandal. The school authorities did not decide to expel the pair; either Principal Medenica was a broad-minded man, or else his modern wife, Varvara, influenced him. On the other hand, neither did the pair give in. And so their love lasted, more or less in the open, until the end of the year, when life separated them.

Everyone admired their resolution, encouraged it, and took pride in it, as something for which they themselves yearned. There was seemingly more to this than the triumph of love.

Truly it was good to love like Mučalica, if only one could find a Ljubica. But there was none to be had.

My love for Dobrica remained sad, unvoiced, and enduring. She must have suspected my love. We always sought one another out later and took joy in our meetings, though we saw each other only once every two or three years. So it was until the war, in which she perished, together with her children. But that was only puppy love.

Real love, with irrepressible passions that surged like underground rivers, was yet to come, with real youth, in the Lim Valley.

First of all, it was Inge. I was sixteen. That love was remote and intangible, nameless but all embracing as a first passion. The entire summer, as I wandered over the hills and cold clear streams, I was haunted by her radiant face, her supple body, those legs, eyes, and distended nostrils of a doe. To be sure, I neither recognized nor admitted that my own yearning was but a part of a desire which gripped the entire school and nearly all the males in town. I must have been partly attracted by her origin and family, which were unusual for that part of the country. Her father was a Serb, a chief judge, and her mother was a German

woman, the same Mrs. Ugrić who taught German. They
had lived long in Germany and when they came the
children knew Serbian badly. There were three sisters and
a brother. However, only the sisters had any importance
for the town. By their way of life, the way they dressed,
and other things as well, they attracted the attention of the
whole town, which was half patriarchal and half peasant.

The Ugrić girls were freer in their conduct. They took
walks alone around town with teachers and officers, and
came home late. The middle sister, Herta, even went horseback riding. They were the first girls to go swimming with
the boys in the Lim. The fashion of the time consisted of
skirts just below the knees; the Ugrić girls wore their skirts
higher than the other girls. But they were also the first to
lengthen them when the style changed. So, too, with their
coiffure. At first they seemed shocking and alluring, and
then everyone began to get enthusiastic and copy them.

Their life in the beginning must have been very difficult.
They were regarded by everybody as rare birds and were
talked about so much that other girls avoided their company. The German girls were doomed to a long isolation
and to the company of men who did not always enjoy
the most sterling reputation. All eyes were fastened on
them, stripped them to their bare skins, examined their
underclothes, inspected their food, followed them through
parks and meadows, and pierced their thoughts and desires.

The eldest, Heide, was a bit more reserved, stout and
plump, with a white-and-pink complexion. She was already
a young lady. It was time to think of marriage, and because she knew Serbian badly as yet, her social life with
young men was hindered, in a small town that watched
every step she made. Finally she married Dragiša Boričić.
The old bachelor succumbed to this beauty, which was
overripe but as cool as a statue. She took marriage seriously.
She set up housekeeping and would not let her husband
budge from her side. Apparently the confirmed and rest-

less bachelor had already begun to grow old and felt only nostalgia for his youth.

The middle sister, Herta, was the sporting type, which was also unusual for the small town. She looked it in her short skirt, with flushed bare arms and legs, and with thick soles on her shoes. Of all the Ugrić girls it was she who aroused the most criticism, for she paid no attention whatever to the clucking of the town but went freely with whom and where she pleased. The youngest, Inge, was in 1926 only a girl, but one who was maturing rapidly. All the unmarried men were beginning to notice her, as though anticipating that those lean limbs would yield a future warm and supple beauty. And when that beauty suddenly burst forth, the whole town was startled, as if at a sudden flash of brilliant light, though they had expected something of the sort. Inge knew herself that she was beautiful, and she acted accordingly. She studied little, encouraged everyone to fall in love with her, but chose no one. My love for her was obviously not much stronger or more lasting than that which others felt for her. I was more aroused and hurt by Labud's unhappy love for her than by my own, and all the more so because I had to break with Labud because of that love. All the less so, however, because I had already, in the beginning of my next-to-last year of high school, fallen in love with another girl, Dušanka, or Duša, who was in the class below mine. That love was not a whit happier than Labud's love for Inge. But neither did it have that depth. This was no longer a boyish enthusiasm, but something that lasted and could not be shaken off at will.

I had just passed sixteen. It was summer, and I suddenly noticed how different everything was—the rocks and the furniture and the doorposts. Everything was alive and imbued with vitality, so that everything I touched or looked upon had something warm coursing through it, like blood. The air, too, all the way to the sky, was filled with the vivacious shimmering of the stars, which stood fixed but radiantly joyful over everything in the inaccessible heights. It was then that I began to lose that horror of apparitions, in which I began to disbelieve. Though there remained forever a certain dread of the dark, now I also found in it a certain warmth and security, and, above all, a sweet solitude, devoid of things or of men, just dreams and the distant remote heavens.

That summer all the surroundings of our house took on a new hue and a certain inner—I cannot say meaning—rather, tension that seemed to seek release in something as yet unknown. Even before this I loved the vast rounded hills and towering peaks, and the Tara, and Jezero, and the brooks which splashed over the pebbles. But now all this took on a transparency through which one could feel that other pulsating life that filled everything.

I had fished for trout and perch before, too, but now this was the stuff dreams are made of, something to be awaited eagerly and to be conducted with a warm yearning. I loved to fish, not only for the cold clear waters and their colors, but for the anticipation of the unknown, that stirring game with nature and her creatures. This was a tie with her inner life, with that warm current.

There was one unforgettable day, hot, so that everything was ablaze and breathed with a beckoning and consuming fire. The ferns were infused with an intoxicating primeval fragrance, the raspberries were bursting with juice, and the swollen streams echoed with a deafening roar which reached the skies. I spent the whole day walking with the shepherdess Mara, as someone enchanted. We were the same age. She moved about, supposedly after her cattle, approaching me and then running away. We said nothing, nor did we touch one another. We did not even look at one another, except for furtive glances that, to me and surely to her as well, seemed crazed and ravaged by an inner fire, the glances of persons devoid of reason. Something had to be said, to be done. But I was paralyzed, tied

into a thousand knots by my own desire. Yet when I left
for a time, when I did not see her, my passion became
quite apparent. It seemed to emanate from everything in all
of space and to focus in me only to seek expression through
me. I felt that despite these endless forests and impassable
chasms I could find this shepherdess just by following her
female scent, without really looking where I was going,
without hearing anything—by the most direct route.

And then it happened one warm and dark evening. A
certain girl—I felt as if I had known her since her birth—
came by on a horse. Suddenly the horse slid down the steep
slope beside a whirlpool in the river. She could not get him
out, though she managed to reach the ford. I swam across
the Tara to extricate the trembling beast. I led the horse
out to the other bank. He was even handsomer now that
he was wet and gleaming. The girl, who was fancily
dressed, thanked me too much and somehow did not seem
to want to leave. Her dark green eyes smiled and beckoned. I trembled, as though from the cold. At last she
leapt easily on her horse. Through the seductive forest,
now without that seductive girl, was heard the soft clatter
of hoofs, which faded and vanished forever. That evening
all of space, up to the starry sky, was filled with the stifling
smell of new-mown hay, which intoxicated me and satu-
rated every cell of my body.

Such happy encounters are like dreams. Yet they cannot be forgotten. Weary and sad, still I felt fortunate that
evening that I had lacked the courage to undertake anything, and that both the shepherdess and the girl, like other
women, had come and gone. For now I perceived all—the
sadness of barren cows, the frenzied neighing of the stallion,
the passion of my older brother which spent itself on
naked flesh, and the quivering beauties of these mountains.

Yet what attracted me most of all was just to sit, on
autumn days, beside the Lim. To keep the swelling river
from overflowing during a torrent, willow sheaves,

weighted down with rocks, were piled along the jagged
banks to the water's edge. I would sit on them for hours,
gazing at the same mass of foam, which lingered over the
rocky ledge like a grazing herd, wondering where to turn.
In the willow sheaf a twig had taken root, and I could see
it suck the earth's strength and grow.

I used to walk to the ancient monastery of St. George's
Columns, which consisted of two parts built in different
periods, not at all beautiful had the monastery not been an
old and historic one. Now I went there more often—and
observed everything. Our rulers knew how to pick the
loveliest spots for their modest memorials. There was a cold
spring below the monastery, and above it a glade of young
oaks, surrounded by gardens and fields. Everything recalled
the past, but also desolation—the moss clustered on the
lintel, the gaping windows of the school the Turks had
razed, the walls inside the monastery stripped of their
frescoes to the bare stone, except for the eyes of some saint
or a fragment of some mighty magnate's vestment. Before
the liberation from the Turks, Mojsije Zečević, a friend of
Bishop Njegoš, was long there as a monk. More of a rebel
than a man of God, he contended with the Turks, striving
to outwit them without losing his head. Many stories remained after him, not only about his heroic exploits, but
how he used to bathe in the icy spring below the monastery
to subdue his desire. Though I saw only ancient ruins, still
I could sense that they contained an unextinguished force
which was capable of awakening and moving again. It was
autumn, and everywhere one felt a young strength which
was withering and yet struggling to survive. So it was
beside the Lim, where the wind bore the smell of damp
fields and ripened gardens, while the roaring Lim brought
from Albania its cold freshness.

Beran Krš is a crumbling cliff that rises out of the plain.
With its yellow grasses against the black rock, its startling
colors and shapes, it seemed like a nocturnal haven for

phantoms or a petrified giant that was falling apart. Withdrawn in the crevices of this cliff, I could hear sounds I
had never heard before—the whispering roots of the reeds,
the cries of the harvesting women in the village below. The
whole earth hummed and trembled, like a glass jar during
the ringing of the bell in the weather-beaten monastery
tower. So, too, were the aching reverberations of Dostoevsky and Strindberg, Nietzsche and the *Kreutzer Sonata*,
the call of the muezzin and the song of a maiden in her
garden.

Everything had ripened just to die, and yet it was resisting and would be born again—the Lim, the monastery,
Beran Krš, the fields and orchards, the changing sky, and
the vague and unbearable yearning of everything for everything.

Perhaps it was a surfeit of reading, since childhood, plus
a tendency to daydream which caused me to experience
love so late, so timidly, and therefore so painfully. Yet
nobody really knows how and why anyone begins to love
anyone else.

It might have been some other girl, for Duša was not
prettier than the others. Plump, with a thick braid in her
hair, a low forehead, and a small pouting mouth, she was a
brunette, with downy cheeks and something gypsy-like
about her, especially those big brilliant eyes which moved
with such a sly languor. During the entire time in Berane
that I was in love with her, I did not exchange a single word
with her. Still, it was love. Everything that had any connection with her—her house, family, even stories of her
heroic forebears—aroused me in the same way. I remembered everything about her: the dresses, movements, and
cooing, caressing voice, and the way she bent over her
desk. I knew everything about her: when she awoke and
went to bed, how well she did at school, and what she
liked. But I did not know her. In the course of nearly the
next two years I did nothing to approach her. She, on the

other hand, was in love with somebody else and maintained a scornful but interested attitude toward me. I wavered ceaselessly between disappointment and hope. I came to believe that I must do something special to attract her. My literary works, which were already being printed, apparently were not enough. I would walk in front of her window as if lost in thought, or even as if a bit drunk. She laughed stupidly at all of this. My half-peasant attire, sandals and coarse homespun, began to torture me. At last I understood why my brother liked to dress like a dandy. Since this was impossible for me (I thought it rather shallow), I felt that I could at least be different.

From the winter of 1929 everything began to get complicated, and a cold bitterness began slowly to crowd out my love. Duša liked Professor Ilija Marković, the only Communist in the school. At that time I was already inclined to Communism and I had to resolve a conflict within myself—between my love for Duša and my philosophical and moral tie with the bearer of an idea to which I was increasingly receptive.

Besides, in addition to that bitterness and dilemma between ideology and feelings, there were my first affairs with my cousin Olga, or Olja, which at first did not seem serious, but which quickly dragged us into drunken frenzies. Duša was my first romantic and ideal love, but Olja was my first passion, the first throat and waist and skin, the first mouth that belonged not only to someone else but to me. In Olja, in that encounter with her, were concentrated all the smells of the meadow, the vibration of things, and the shimmering of space. But in that insatiable striving with her there was also something tired and lukewarm, an ebbing even before our passions reached their greatest strength and fullness.

The two of us never admitted or expressed our love. We did not dare, being close relatives without any prospect of a lasting love or marriage. The more we yearned for one another, the more we had to conceal our love. This love was pure of anything save the bare desires of youth, not quite savage, but unquenchable and direct. Olja was of my age, but more experienced than I. She lived the secluded life of maidens in a Moslem town who, precisely because of being shut in, ripen earlier and seek to satisfy their desire, most frequently with kinsmen. She, too, knew all about those petty peccadilloes and naughty pleasures, learned from Lord knows whom, which were passed down from generation to generation. She had just been jilted by a youth whom she liked. But she knew how to get over the hurt quickly, having learned from childhood that loves come and go, but sorrow and desire are inevitable.

There was, however, in that love of ours something that was just ours, something different from similar affairs between cousins, something constant, always yearning to be fulfilled—a dream come true and which lasted between us.

I was torn and rent asunder by those two loves, which were so different in everything. These were the puppy loves of schooldays. Real love, the kind with no duality, which embraces the whole gamut of a man's feelings and thoughts, all his strength—such a love was yet to come.

Dušanka intended to marry Marković, but he died of tuberculosis. At that time I was already at the university. A year later she spoke to me—the first words we ever exchanged. We began a relationship that we believed, in vain, could become something more than friendship. The love I once wanted from her she had already given to another. Now there were for me only mementos of love, beautiful and painful, but dead. This girl, nevertheless, had something great in her, which I found out only much later. Though our liaison burned itself out quickly and was already almost completely dead, she did not wish to marry the whole time I was in prison. During the three years of my imprisonment she openly proclaimed our closeness, like a betrothed in ancient poems. This gave courage to our comrades, and they took pride in her. Only then did I begin to understand: having promised herself to Marković, who was much older than she, she remained true to his cause. Love was for her a great moral obligation.

Dušanka maintained this constancy and pride even later. She married a man of turbulent nature and heroism, someone considerably older than herself and a friend of her former fiancé. The Chetniks killed him. Nevertheless, in front of both them and the occupation forces she proudly proclaimed that she was his wife, and pluckily raised her children by her own efforts, without even thinking of marrying again. This determination stayed with her throughout all the misfortunes that assailed her. And there were more to come, even harder ones. She spent several years in a camp because she sided with the Soviet Central Committee against the Yugoslav. Her long imprisonment was a testimonial to her firmness. She spared neither her children nor herself, for a senseless and unjust cause, but one of which she was convinced.

Who could have suspected that such an indomitable moral force hid behind her once shining and mischievous eyes? Those eyes today have grown bleary with suffering and adversity but also more steadfast and calm.

Olja also got married, while I was a university student. She was one of the prettiest girls in town. Being poor, however, she married an ugly old man. Their marriage was a good one, with many children. Unlike Dušanka, she never bothered about politics; she was completely absorbed by her home and family. Her husband was executed as a leading Chetnik. She emerged from the war worn from misery and suffering, already gray and wrinkled. Only her eyes were still young and pretty, but bereft of that once unrestrained inner glow. She was glad to see me, although I belonged to the side that had killed her husband. She never understood political movements, seeking only peace and happiness in life.

Though I matured fast intellectually and chose an ideology too soon, in love I was late. There was certainly some cause-and-effect relationship there. But I did not know it then.

I wished to fight, even if it meant rejecting love.

A man is not formed all at once but, being a whole, grows first in one way, then in another.

Every man, especially a youth, yearns after various paths in life, and frequently he is forced to take the very one he never quite felt to be his own.

I was the only one of my schoolmates who quite definitely regarded himself as a Communist, even in the eighth grade. But I wished to be a writer. Finding myself even then, and especially later, with the dilemma of choosing between my personal desire and those moral obligations that I felt I owed society, I always decided in favor of the latter. Of course, such a decision is a pleasant self-deception: every man wishes to portray his role in society in the best possible light and as the result of great personal sacrifice and inner dramas. Yet it is true, even where this is so, that a man who rejects self through a struggle nevertheless does only what he has to do, conditioned by the circumstances in which he finds himself and by his own personal traits.

It was neither Marxist literature nor the Communist movement which revealed to me the path of Communism, for neither the one nor the other existed in the backward and primitive environment of Berane.

There lived in town a Communist—the brother of a merchant, an agent for the Singer Company. They called him Singer, too. The very fact that he lived in eccentric solitude and read a lot was enough to draw suspicious attention on him, though he was not active in any way. When

I tried to approach him, he seemed to become frightened, and though he promised to give me something to read, he never did. He was the town wonder, but dead and powerless, like a fountain without water. Later, when a Communist organization was formed in Berane, it ran afoul of the passivity and exaggerated caution of this man, who believed that it was wiser to do nothing illegal, and that it was sufficient to meet legally and to talk. He was, of course, an opportunist and a liquidator, and was rejected and crushed.

Ilija Marković, who came as an instructor in 1926, attracted me most of all, even though he was not an open Communist. He might have been between thirty-five and forty years old. He was gaunt, tall, with an unwholesomely flushed face, curled lips, and bad teeth, extremely large beautiful dark eyes, and a high, tranquil and thoughtful forehead. He was gentle and yielding with the girl pupils, he lectured well, though somewhat carelessly, and he conducted examinations in the same fashion. He engaged in few friendships, but read a good deal. He gave no sign whatever of any organized Communist activity. He did not engage in any. His whole activity consisted of intimate conversations. As a university student he had belonged to a Communist organization and had been active. He was one of that generation of Communists which replaced the first, postwar one, and which developed its own character, neither too militant nor quite conciliatory, in the semilegal circumstances of the dictatorship.

Neither Ilija Marković nor Singer influenced my own development in any decisive way. They did not even enter into conversations about Communism with me. I was too young and too inexperienced for them, and perhaps they were afraid to engage in such dangerous conversations at a time when the royal dictatorship held sway with a severity that found less reason in popular resistance than in its own lack of self-confidence, from which arose its determination to establish itself firmly and to frighten its powerful opponents from the very beginning.

The dictatorship of January 6 suppressed all political and even intellectual activity. It was an exceptional event in my life as well. For many it came as a cataclysmic earthquake. All their established conceptions and ways were shattered. I was then, in the eighth grade, an eyewitness to the submission and withdrawal of people, as though the frigid darkness of midwinter had fettered men's souls. Most people approved the King's action, though irresolutely: the state and unity of the country had been saved, and Parliament had been incapable anyway. Apart from this public approbation, however, there reigned a cold silence which obviously angered those who had eagerly greeted the dictatorship, and there were such even among the teachers, particularly those who had not finished their degrees and who hoped to gain a sure livelihood by their vociferousness. Hitherto indiscernible men now rose to the top—tavern keepers, police clerks, village scalawags, and men generally ready for any desperate deed for the sake of politics and personal ambition.

The district chief, a peaceful and unassuming man, became the most important figure, despite his own wishes. Everybody began to bow to him, to greet him, and to fawn on him, even on the street, in front of others. Till then few people knew or even noticed him. He, on the other hand, assumed the role of a generous man who had understanding for everybody and, moreover, for everything.

The dictatorship did not alter life fundamentally, at least not as I understood the state of affairs: the government, the real one—of gendarmes and policemen—had been ruthless and all-important even before. But before, people had not been afraid to talk, and in the towns they had not greatly feared the police; their elected representatives were of some account. What the dictatorship did was to make an end of free speech, which most people apparently value as much as bread. It did not even touch the material life of the people, at least so it seemed. Indeed, at first it even seemed that things might be better and certainly simpler, without sterile party feuds and elections that decide nothing. The majority, at least in my part of the country, greeted the dictatorship with relief, despite general misgivings. It was the educated men who regarded it with the greatest suspicion and secret dissatisfaction, especially those who had hitherto been active in political parties.

No changes were apparent in Ilija Marković. Our other Communist, Singer, apparently never went out of his house, but waited, for the spring.

With me, at first and also later, the dictatorship only intensified my somber state of mind and discontent. It was the cause both of my spiritual wanderings and of my dissatisfaction with social conditions. It seems to me that it was precisely these repressed dark moods, this psychologizing, that provided the base for a political and social discontent which was all the more profound because it was moot and unconscious—out of the very fabric of the soul, out of every pore of one's inner life. Later, in Belgrade, when I became acquainted with my fellow students at the university, I noticed that they, too, each in his own way, to be sure, had traveled the same path—the same literature à la Dostoevsky and Krleža,* the same inner crises and somber moods, dark discontent and bitterness over cruelty and injustice among men and in society generally. Hence also a certain contrived, concocted attitude, rather pretentious and no less disheveled, unstrung, and rebellious.

It was classical and humanistic literature that drew me to Communism. True, it did not lead directly to Communism, but taught more humane and just relations among men. Existing society, and particularly the political movements within it, were incapable even of promising this.

---

* Krleža Miroslav, a Croat, was a novelist and Marxist writer who had a great influence on young Yugoslav intellectuals between the two World Wars and oriented many of them toward Communism.

At that time I was reading Chernyshevsky.* He and his clumsy novel could not make any particular impression on me, certainly because it was so completely unconvincing and shallow as a literary work. He might have been able to rear a series of revolutionary generations in Russia, and to have a significant influence even in our country until modern times, but for the generation under the dictatorship he was without any significance. Such utopian musings, sentimental stories, and the like left no lasting traces. *Uncle Tom's Cabin* and Hugo's *Les Miserables* caused only a temporary impress, albeit a very strong one, which was forgotten when the book was laid aside. Marxist or socialist literature of any kind did not exist at all in Berane at the time, nor was it to be had. The only thing that could exert any influence, and indeed did, was great literature, particularly the Russian classics. Its influence was indirect, but more lasting. Awakening noble thoughts, it confronted the reader with the cruelties and injustices of the existing order.

Yet it was the state of society itself that provided the prime and most powerful impulse. If anyone wished to change it—and there are always men with such irresistible desires—he could do so only in a movement that promised something of the kind and was said to have succeeded once through a great revolution. The guardians of the *status quo* only made something like this attractive to a young man by their stories about the Communist specter and by their panicky preservation of old forms and relations.

This was a desire for a better and happier life, for change, which is inborn in every creature, and which in certain concrete conditions could not take on any but the Communist form. Communism was a new idea. It offered youth enthusiasm, a desire for endeavor and sacrifice to achieve the happiness of the human race.

* Chernyshevsky(i) was a Russian revolutionary who died in exile in Siberia in 1889 and whose work *What Is to Be Done* (1863) was accounted a classic by later Russian radicals.

Ilija Marković knew that I felt drawn to Communism, and that I was in love with a girl toward whom he had more serious intentions than those of a high-school student. I could tell by the kind and considerate way he treated me. This would have been gratifying, for it showed his generosity, had it not struck me as being a contrived pity, which I had never asked from anyone.

I was beset by questions that shook all of my previous moral, emotional, and intellectual conceptions. Was it honest for an older man, moreover a sick man, to entice—even with intentions of marriage—a girl of sixteen or seventeen years? And his pupil at that! True, there had already been marriages between instructors and pupils in the school. But such things were not done by the bearers of such great ideas as Communism, which was supposed to bring not only justice and an end to misery, but a new morality among men.

And what was I to do, if that is how it was? Was I to love or to hate this man? Was I to hold him in contempt or to admire him as a contender for the same ideal? The posing of these and similar questions did not at all affect actual relationships, but had vast importance for my inner life and further development. On the answer depended the growth of my inner moral personality. Of course, I answered straight away: There is no real reason to hate him; this would be selfish and unmanly on my part. Yet from this answer to a corresponding reality within myself there was a very long and painful path full of mental twists and turns and visions that could only excite moral repugnance and even jealousy. Feeling that I hated this man, I suppressed the hate.

I succeeded even in liking him, though without warmth, even more than was required by our tie, either personal or ideological. Through this I got over my love. That inner metamorphosis, which ended in my stifling within myself both jealousy and love, quickened my vague progress toward Communism and conscious turning to literature.

It was as though my adherence to Communism, too, depended on my success in mastering myself in this personal morality play. This was my first great sacrifice, in the name of nobility, even a pretended one.

My last year in high school was full of painful and complex inner conflicts. This was followed, finally, by a certain clarity, at least in the form of the question to write or to fight. Even then, future lines and tendencies made their appearance and left their mark in the midst of troubled psychological conflicts, social discontent, and an overtaxing nostalgia. From this moral and emotional crisis I emerged strengthened, with some bitterness inside myself, but with an ethical principle—that one should not hate men for personal reasons, and that one should not mix personal needs and problems with one's ideology.

At the end of the school year, on St. Vitus's Day, the majority of my schoolmates appeared with canes and ties. These were considered the signs of maturity of the graduating students. It all seemed to me too common and formalistic. I also put on a tie, but a different kind—a red one. I thought about it a long time before I did it, for a tie of that color was the badge of a Communist, and none dared to wear it. If I am a Communist, I thought, and I am, then I must be publicly true to that conviction. There was childish bravado in this, but also defiance at a time and in a place where no one was defiant.

There existed—and perhaps still exists—a picture of myself just after graduation, in a Russian-style peasant shirt and a sash, with my arm hanging over the back of a chair. I had recalled even before sitting in front of the camera, that Tolstoy held his arm in the same way in a certain picture and that he, too, wore a peasant shirt. I was consciously imitating him. The shirt—its cut and everything —was designed by me in imitation of Tolstoy, of the Russians. Later it caught my fancy, both for its originality and practicality, and I wore it as a university student. Despite such imitations of every kind, which I carried to an extreme, there was then, and after, and in these very imitations, a dark inner turbulence, a profound dissatisfaction with the existing state of affairs and with the limitations they placed on human and social potentialities. A vague inner spiritual and intellectual torment beset me even then and would not let me go.

Marković came up to me after the diplomas were presented, obviously as man to man because of Dušanka, and as comrade to comrade because of Communism. He walked with me from the school to town, telling me, sagely and gently, how in Belgrade, at the university, everything would be nicer and better: many new friends, a life of greater ease, a more progressive and developed environment. But there was no need either to console or to encourage me. I had already made my peace with many things —with sentimental love and with helping the world through charity. Things and human relations presented themselves in ever harder and harsher forms. It was still a land without justice.

I spent the summer in Bijelo Polje, where my family had already resettled. Bijelo Polje was similar to Berane in many ways, except that the Moslem population in and around it was more numerous. Its way of life was still patriarchal, its houses poorer, and the uncleanliness even greater. There was not even a dirt road to connect it with any other town except Plevlje. Here was a remote region, rich in fruit that rotted away unused, godless, filled with the halt and the blind. The rebellious and overweening Vasojević tribe had poured into the Lim plain and had taken over both it and the little town of Berane. Here, however, the Montenegrins were interlopers who had forced their way into a town that was not theirs. The former Turkish landlords of Berane were hardly noticed, but here their adversity filled every little corner of life—

their songs and stories, evening gatherings under the old pear trees, and the desperate nightly carousing.

But this did not concern me then. I was preparing myself for a new world, with my eyes already opened to comprehend it and with a troubled soul, fearful of becoming lost in it.

INDEX